... he did as a twenty-one-year-old graduate student. His 1989 book, *Longing for the Harmonies*, was a *New York Times* Notable Book of the Year. He lives in Cambridge, Massachusetts, where he is currently the Herman Feshbach Professor of Physics at MIT.

THE LIGHTNESS OF BEING

Big Questions, Real Answers

Frank Wilczek

PENGUIN BOOKS

PENGUIN BOOKS

Published by the Penguin Group
Penguin Books Ltd, 80 Strand, London WC2R 0RL, England
Penguin Group (USA) Inc., 375 Hudson Street, New York, New York 10014, USA
Penguin Group (Canada), 90 Eglinton Avenue East, Suite 700, Toronto, Ontario, Canada M4P 2Y3
(a division of Pearson Penguin Canada Inc.)
Penguin Ireland, 25 St Stephen's Green, Dublin 2, Ireland (a division of Penguin Books Ltd)
Penguin Group (Australia), 250 Camberwell Road, Camberwell, Victoria 3124, Australia
(a division of Pearson Australia Group Pty Ltd)
Penguin Books India Pvt Ltd, 11 Community Centre, Panchsheel Park, New Delhi – 110 017, India
Penguin Group (NZ), 67 Apollo Drive, Rosedale, North Shore 0632, New Zealand
(a division of Pearson New Zealand Ltd)
Penguin Books (South Africa) (Pty) Ltd, 24 Sturdee Avenue, Rosebank, Johannesburg 2196, South Africa

Penguin Books Ltd, Registered Offices: 80 Strand, London WC2R 0RL, England

www.penguin.com

First published in the United States of America by Basic Books,
a Member of the Perseus Books Group 2008
First published in Great Britain by Allen Lane 2009
Published in Penguin Books 2010
1

Printed in England by Clays Ltd, St Ives plc

978-0-141-04314-2

Dedicated to the memory of

Sam Treiman and Sidney Coleman—

guides in science, friends in life.

Contents

About the Title

The Unbearable Lightness of Being is the title of a famous novel by Milan Kundera—one of my favorite books. It is about many things, but perhaps above all the struggle to find pattern and meaning in the seemingly random, strange, and sometimes cruel world we live in. Of course Kundera's approach to these problems, through story and art, looks very different from the one I've taken in this book, through science and (light) philosophy. For me at least, coming to understand the deep structure of reality has helped to make Being seem not merely bearable, but enchanted—and enchanting. Hence, *The ~~Unbearable~~ Lightness of Being.*

There's also a joke involved. A central theme of this book is that the ancient contrast between celestial light and earthy matter has been transcended. In modern physics, there's only one thing, and it's more like the traditional idea of light than the traditional idea of matter. Hence, *The* ***Lightness*** *of Being.*

Reader's Guide

THE CORE PLAN OF THIS BOOK couldn't be simpler: it is meant to be read chapter by chapter, from beginning to end. But I've also supplied:

- An extensive glossary, so that you don't get tripped up by unfamiliar words or have to search for the place fifty pages back where they were introduced. It can also be mined for cocktail party nuggets. It's even got a few jokes.
- Endnotes that elaborate on fine points, follow some important tangents, or provide references.
- Three appendices. The first two take discussions in Chapters 3 and 8, respectively, into deeper waters; the third is a first-person account of how a key discovery reported in Chapter 20 occurred.
- A web page, itsfrombits.com, where you'll find additional pictures, links, and news related to the book.

You can detour to the appendices as their chapters come up, but if you prefer to get on with the story instead, you should still find it comprehensible. I considered offloading more of the material in Chapter 8, but in the end I couldn't bring myself to do it. So in that chapter you'll find *much* ado about Nothing.

PART I

The Origin of Mass

MATTER IS NOT WHAT IT APPEARS TO BE. Its most obvious property—variously called resistance to motion, inertia, or mass—can be understood more deeply in completely different terms. The mass of ordinary matter is the embodied energy of more basic building blocks, themselves lacking mass. Nor is space what it appears to be. What appears to our eyes as empty space is revealed to our minds as a complex medium full of spontaneous activity.

1

Getting to *It*

The universe is not what it used to be, nor what it appears to be.

WHAT'S IT ALL ABOUT? People reflecting upon the wide world around them, the varied and often bewildering experience of life, and the prospect of death are driven to ask that question. We seek answers from many sources: ancient texts and continuing traditions, the love and wisdom of other people, the creative products of music and art. Each of these sources has something to offer.

Logically, however, the first step in the search for answers should be to understand what "it" is. Our world has some important and surprising things to say for itself. That's what this book is about. I want to enrich your understanding of just what "it" is that you and I find ourselves within.

Senses and World-Models

To begin, we build our world-models from strange raw materials: signal-processing tools "designed" by evolution to filter a universe swarming with information into a very few streams of incoming data.

Data streams? Their more familiar names are vision, hearing, smell, and so forth. From a modern point of view, vision is what samples the electromagnetic radiation that passes through tiny holes in our eyes, picking up only a narrow rainbow of colors inside a much broader spectrum. Our hearing monitors air pressure at our eardrums, and smell provides a quirky chemical analysis of the air impinging on our nasal membranes. Other sensory systems give some rough information about the overall acceleration of our body (kinesthetic sense), temperatures and pressures over its surface (touch), a handful of crude measures of the chemical composition of matter on our tongue (taste), and a few other odds and ends.

Those sensory systems allowed our ancestors—just as they allow us—to construct a rich, dynamic model of the world, enabling them to respond effectively. The most important components of that world-model are more-or-less stable objects (such as other people, animals, plants, rocks, . . . the Sun, stars, clouds, . . .) some of them moving around, some dangerous, some good to eat, and others—a select and especially interesting few—desirable mates.

Devices to enhance our senses reveal a richer world. When Antonie van Leeuwenhoek looked at the living world through the first good microscopes in the 1670s, he saw totally unsuspected, hidden orders of being. In short order he discovered bacteria, spermatozoa, and the banded structure of muscle fibers. Today we trace the origin of many diseases (and of many benefits) to bacteria. The basis of heredity (well, half of it) is found within the tiny spermatozoa. And our ability to move is anchored in those bands. Likewise, when Galileo Galilei first turned a telescope to the sky in the 1610s, new riches appeared: he found spots on the Sun, mountains on the Moon, moons around Jupiter, and multitudes of stars in the Milky Way.

But the ultimate sense-enhancing device is a thinking mind. Thinking minds allow us to realize that the world contains much more, and is in many ways a different thing, than meets the eye. Many key facts about the world don't jump out to our senses. The parade of seasons, in lock-step with the yearly cycle of sunrise and sunset, the nightly rotation of stars across the sky, the more intricate but still predictable motions of the Moon and planets, and their connection with eclipses—these patterns do not leap to the eye, ear, or nose. But thinking minds can discern them. And having noticed those regularities, thinking minds soon discover that they are *more* regular than the rules of thumb that guide our everyday plans and expectations. The more profound, hidden regularities lend themselves to counting and to geometry: in short, to mathematical precision.

Other hidden regularities emerged from the practice of technology—and, remarkably, of art. The design of stringed musical instruments is a beautiful and historically important example. Around 600 BCE, Pythagoras observed that the tones of a lyre sound most harmonious when the ratio of string lengths forms a simple whole-number fraction. Inspired by such hints, Pythagoras and his followers made a remarkable intuitive leap. They foresaw the possibility of a different kind of world-model, less dependent on the accident of our senses but more in tune with Nature's hidden harmonies, and ultimately more faithful to reality. That is the meaning of the Pythagorean Brotherhood's credo: "All things are number."

The scientific revolution of the seventeenth century began to validate those dreams of ancient Greece. That revolution led to Isaac Newton's mathematical laws of motion and of gravity. Newton's laws permitted precise calculation of the motion of planets and comets, and provided powerful tools for describing the motion of matter in general.

Yet the Newtonian laws operate in a world-model that is very different from everyday intuition. Because Newtonian space is infinite and homogeneous, Earth and its surface are not special places. The directions "up," "down," and "sideways" are fundamentally similar. Nor is rest privileged over uniform motion. None of these concepts matches everyday experience. They troubled Newton's contemporaries, and even Newton himself. (He was unhappy with the relativity of motion, even though it is a logical consequence of his equations, and to escape it he postulated the existence of "absolute" space, with respect to which true rest and motion are defined.)

Another big advance came in the nineteenth century, with James Clerk Maxwell's equations for electricity and magnetism. The new equations captured a wider range of phenomena, including both previously known and newly predicted kinds of light (what we now call ultraviolet radiation and radio waves, for example), in a precise mathematical world-model. Again, however, the big advance required a readjustment and vast expansion of our perception of reality. Where Newton described the motion of particles influenced by gravity, Maxwell's equations filled space with the play of "fields" or "ethers." According to Maxwell, what our senses perceive as empty space is actually the home of invisible electric and magnetic fields, which exert forces on the matter we observe. Although they begin as mathematical devices, the fields leap out of the equations to take on a life of their own. Changing electric fields produce magnetic fields, and changing magnetic fields produce electric fields. Thus these fields can animate one another in turn, giving birth to self-reproducing disturbances that travel at the speed of light. Ever since Maxwell, we understand that these disturbances are what light *is*.

These discoveries of Newton, Maxwell, and many other brilliant people greatly expanded human imagination. But it's only in twentieth and twenty-first century physics that the dreams of Pythagoras truly approach fruition. As our description of fundamental processes becomes more complete we see more, and we see differently. The deep structure of the world is quite different from its surface structure. The senses we are born with are not attuned to our most complete and accurate world-models. I invite you to expand your view of reality.

Power, Meaning, and Method

When I was growing up, I loved the idea that great powers and secret meanings lurk behind the appearance of things.[1] I was

1. I still do!

entranced by magic shows and wanted to become a magician. But my first magic kit was a profound disappointment. The secret of the magic, I learned, was not genuine power, just trickery.

Later, I was fascinated by religion: specifically, the Roman Catholic faith in which I grew up. Here I was informed that there are secret meanings behind the appearance of things, great powers that can be swayed by prayer and ritual. But as I learned more about science, some of the concepts and explanations in the ancient sacred texts came to seem clearly wrong; and as I learned more about history and historiography (the recording of history), some of the stories in those texts came to seem very doubtful.

What I found most disillusioning, however, was not that the sacred texts contained errors, but that they suffered by comparison. Compared to what I was learning in science, they offered few truly surprising and powerful insights. Where was there a vision to compete with the concepts of infinite space, of vast expanses of time, of distant stars that rival and surpass our Sun? Of hidden forces and new, invisible forms of "light"? Or of tremendous energies that humans could, by understanding *natural* processes, learn to liberate and control? I came to think that if God exists, He (or She, or They, or It) did a much more impressive job revealing Himself in the world than in the old books—and that the power of faith and prayer is elusive and unreliable compared to the everyday miracles of medicine and technology.

"Ah," I hear the traditional believer object, "but scientific study of the natural world does not reveal its *meaning*." To which I reply: Give it a chance. Science reveals some very surprising things about what the world is. Should you expect to understand what it means, before you know what it is?

In Galileo's time, professors of philosophy and theology—the subjects were inseparable—produced grand discourses on the nature of reality, the structure of the universe, and the way the world works, all based on sophisticated metaphysical arguments. Meanwhile, Galileo measured how fast balls roll down inclined planes. How mundane! But the learned discourses, while grand, were vague. Galileo's investigations were clear and precise. The old metaphysics never progressed, while Galileo's work bore abundant, and at length spectacular, fruit. Galileo too cared about the big questions, but he realized that getting genuine answers requires patience and humility before the facts.

That lesson remains valid and relevant today. The best way to address the big ultimate questions is likely to be through dialogue with Nature. We must pose pointed sub-questions that give Nature a chance to respond with meaningful answers, in particular with answers that might surprise us.

This approach does not come naturally. In the life we evolved for, important decisions had to be made quickly using the information at hand. People had to spear their prey before they *became* the prey. They could not pause to study the laws of motion, the aerodynamics of spears, and how to compute a trajectory. And big surprises were definitely *not* welcome. We evolved to be good at learning and using rules of thumb, not at searching for ultimate causes and making fine distinctions. Still less did we evolve to spin out the long chains of calculation that connect fundamental laws to observable consequences. Computers are much better at it!

To benefit fully from our dialogue with Nature, we must agree to use Her language. The modes of thought that helped us to survive and reproduce on the African savannah of 200000 BCE will not suffice. I invite you to expand the way you think.

The Centrality of Mass

In this book we'll explore some of the grandest questions imaginable: questions about the ultimate structure of physical reality, the nature of space, the contents of the Universe, and the future of human inquiry. Inspired by Galileo, however, I will address these questions as they arise in the course of a natural dialogue with Nature, about a specific topic.

The topic that will be our doorway into much bigger questions is *mass*. To understand mass deeply, we'll move past Newton, Maxwell, and Einstein, calling on many of the newest and strangest ideas of physics. And we'll find that understanding mass allows us to address very fundamental issues about unification and gravity that are at the forefront of current research.

Why is mass so central? Let me tell you a story.

Once upon a time there was something called matter that was substantial, weighty, and permanent. And something else, quite different, called light. People sensed them in separate data streams; touching one, seeing the other. Matter and light served—and still do serve—as powerful metaphors for other contrasting aspects of reality: flesh and spirit, being and becoming, earthy and celestial.

When matter appeared from nowhere, it was a sure sign of the miraculous, as when Jesus served the multitude from six loaves of bread.

The scientific soul of matter, its irreducible essence, was mass. Mass defined matter's resistance to motion, its inertia. Mass was unchangeable, "conserved." It could be transferred from one body to another but could never be gained or lost. For Newton, mass *defined* quantity of matter. In Newton's physics, mass provided the link between force and motion, and it provided the source of gravity. For Lavoisier, the persistence of mass, its accurate conservation, provided the foundation of chemistry, and offered a fruitful guide to discovery. If mass seems to disappear, look for it in new forms—*voilà*, oxygen!

Light had no mass. Light moved from source to receptor incredibly fast, without being pushed. Light could be created (emitted) or destroyed (absorbed) very easily. Light exerted no gravitational pull. And it found no place in the periodic table, which codified the building blocks of matter.

For many centuries before modern science, and for the first two and a half centuries *of* modern science, the division of reality into matter and light seemed self-evident. Matter had mass, light had no mass; and mass was conserved. As long as the separation between the massive and the massless persisted, a unified description of the physical world could not be achieved.

In the first part of the twentieth century, the upheavals of relativity and (especially) quantum theory shattered the foundations beneath classical physics. Existing theories of matter and light were reduced to rubble. That process of creative destruction made it possible to construct, over the second part of the twentieth century, a new and deeper theory of matter/light that removed the ancient separation. The new theory sees a world based on a multiplicity of space-filling ethers, a totality I call the Grid. The new world-model is extremely strange, but also extremely successful and accurate.

The new world-model gives us a fundamentally new understanding of the origin of the mass of ordinary matter. How new? Our mass emerges, as we'll discuss, from a recipe involving relativity, quantum field theory, and chromodynamics—the specific laws governing the behavior of quarks and gluons. You *cannot* understand the origin of mass without profound use of all these concepts. But they all emerged only in the twentieth century, and only (special) relativity is really a mature subject. Quantum field theory and chromodynamics remain active areas of research, with many open questions.

High on their success, and having learned much from it, physicists enter the twenty-first century with ideas for further syntheses. Today, ideas that go far toward achieving a unified description of the superficially different forces of nature, and toward achieving a unified account of the superficially different ethers we use today, are ready for testing. We have some subtle, tantalizing hints that those ideas are on the right track. The next few years will be their time of trial, as the great accelerator LHC (Large Hadron Collider) begins to operate.

listen: there's a hell of a good universe next door; let's go.
— e e cummings

2

Newton's Zeroth Law

What is matter? Newtonian physics supplied a profound answer to that question: matter is that which has mass. While we no longer see mass as the ultimate property of matter, it is an important aspect of reality, to which we must do justice.

IN *MATHEMATICAL PRINCIPLES OF NATURAL PHILOSOPHY* (1686), the monumental work that perfected classical mechanics and sparked the Enlightenment, Isaac Newton formulated three laws of motion. To this day, courses on classical mechanics usually begin with some version of Newton's three laws. But these laws are not complete. There is another principle, without which Newton's three laws lose most of their power. That hidden principle was so basic to Newton's view of the physical world that he took it not as a law that governs the motion of matter, but as the *definition* of what matter *is*.

When I teach classical mechanics, I start by bringing out the hidden assumption I call Newton's zeroth law. And I emphasize that it is wrong! How can a definition be wrong? And how can a wrong definition be the foundation for great scientific work?

The legendary Danish physicist Niels Bohr distinguished two kinds of truths. An ordinary truth is a statement whose opposite is a falsehood. A profound truth is a statement whose opposite is also a profound truth.

In that spirit, we might say that an ordinary mistake is one that leads to a dead end, while a profound mistake is one that leads to progress. Anyone can make an ordinary mistake, but it takes a genius to make a profound mistake.

Newton's zeroth law was a *profound* mistake. It was the central dogma of an Old Regime that governed physics, chemistry, and astronomy for more than two centuries. Only at the beginning of the twentieth century did the work of Planck, Einstein, and others begin to challenge the Old Regime. By mid-century, under bombardment from new experimental discoveries, the Old Regime had crumbled.

That destruction opened the way to a new creation. Our New Regime frames an entirely new understanding of what matter is. The New Regime is based on laws that differ from the old ones, not merely in detail but also in kind. This revolution in basic understanding, and its consequences, are what we'll be exploring.

But to justify the revolution we must first bring the shortcomings of the Old Regime into clear focus. For its mistakes are, in Bohr's sense, profound. The Old Regime of Newtonian physics gave us relatively simple and easy-to-use rules with which we could govern the physical world pretty effectively. In practice, we still use those rules to administer the more peaceful, well-settled districts of reality.

So, to begin, let's take a close look at Newton's hidden assumption, his zeroth law—both its tremendous strength and its fatal weakness. That law states that mass is neither created nor destroyed. Whatever happens—collisions, explosions, a million years of wind and rain—if you add up the total mass of all the material involved at the beginning, or at the end, or at any intermediate time, you will always get the same sum. The scientific jargon for this is that mass is conserved. The standard, dignified name for Newton's zeroth law is *conservation of mass.*

God and the Zeroth Law

Of course, to translate the zeroth law into a meaningful, scientific statement about the physical world, we have to specify how masses are measured and compared. We'll do that momentarily. But let me first highlight why the zeroth law is not just another scientific law, but a strategy for understanding the world—a strategy that looked very good for a very long time.

It's revealing that Newton himself usually used the phrase *quantity of matter* for what we now call mass. His wording implies that you can't have matter without mass. Mass is the ultimate measure of matter; it tells you how much matter you've got. No mass, no matter. Thus the conservation of mass expresses—indeed, is equivalent to—the persistence of matter. For Newton, the zeroth law was not so much an empirical observation or experimental discovery as a necessary truth; it was not a proper law at all, but a definition. Or rather, as we'll see in a moment, it expressed a religious truth—a fact about God's method of creation. (To avoid misunderstanding, let me emphasize that Newton was a meticulous empirical scientist, and he carefully checked that the consequences of his definitions and assumptions described Nature as accurately as the measurements of the day could test them. I'm not saying that he let his religious ideas trump reality. It's more subtle: those ideas gave him his intuition about how reality works. What motivated Newton to suspect that something like the zeroth law had to be true was not painstaking experiments but, rather, powerful intuition, derived from his religion, about how the world is built. Newton had no doubt about God's existence, and he saw his task in science as revealing God's method of governing the physical world.)

In his later *Opticks* (1704), Newton was more specific in expressing his vision of the ultimate nature of matter:

> It seems probable to me, that God in the beginning formed matter in solid, massy, hard, impenetrable, moveable particles, of such sizes and figures, and with such other properties, and in such proportions to space, as most conduced to the ends for which He formed them; and that these primitive particles being solids, are incomparably harder than any porous bodies compounded of them, even so very hard, as never to wear or break in pieces; no ordinary power being able to divide what God Himself made one in the first creation.

This remarkable passage contains a few points we should notice. First: Newton takes the property of having a fixed mass as one of the most basic properties of the ultimate building blocks of matter. He calls it being "massy." Mass, for Newton, is not something you should try to explain in terms of something simpler. It is part of the ultimate description of matter; it reaches bottom. Second: Newton ascribes the changes we observe in the world entirely to *rearrangements* of elementary building blocks, elementary particles. The building blocks themselves are neither created nor destroyed—they just move around. Once God has made them, their properties, including their mass, never change. Newton's zeroth law of motion, the conservation of mass, follows from those two points.

Getting Real

Now we must return from these heady philosophico-theological ideas about why conservation of mass might be true, or must be true, to the ordinary business of measuring to see whether it *is* true.

How do we measure mass? The most familiar way is to use a scale. One sort of scale, the kind dieters have in their bathrooms, compares how much bodies (that is, dieters' bodies) can compress a spring. Closely related are the scales that anglers use, which compare how much dangling bodies (that is, fish) stretch a spring. The amount the spring stretches (or, for the dieter, compresses) is proportional to the downward force the body exerts, which is what we call the body's weight, which is proportional to its mass.

In this very concrete and practical framework, conservation of mass simply says that a closed system will continue to stretch a spring by the same amount, whatever is going on inside. This is precisely what Antoine Lavoisier (1743–1784) verified—using, to be sure, more sophisticated and accurate scales than you'll find in your bathroom—in the many painstaking experiments that earned him the title "father of modern chemistry." Lavoisier checked, in a wide variety of chemical reactions, that the total weight of all the stuff you started with was equal to the total weight of what you had after the reaction took place, within the accuracy he could measure (typically one part in a thousand or so). Through the discipline of accounting for *all* the matter in a reaction—capturing the gases that might escape, collecting the ashes of explosions, and so forth—he discovered new compounds and elements. Lavoisier was guillotined during the French Revolution. The mathematician Joseph Lagrange said, "It took them only a moment to cut off that head, but France may not produce another like it in a century."

Using scales to compare masses is practical and effective, but it won't do as a general, principled definition of mass. For example, if you take your body out into space, its weight as measured by a scale will get smaller, but its mass will stay the same. (Scales will lie, but waistlines won't shrink.) Mass had better stay the same, if the law of conservation of mass is going to be true! And that superficially circular assertion has real content, because you can compare masses in other ways. For example, you can compare how fast two cannonballs start to fly after you launch them out of the same cannon. According to Newton's other laws of motion, a given impulse will give rise to a velocity inversely proportional to the mass. So if one cannonball comes out twice as fast as the other, it has half the mass—whether you do the experiment at the surface of Earth or in space.

I won't go further into the technicalities of measuring mass, except to say that there are many ways to do it besides using scales and shooting things from cannons, and many checks of their mutual consistency.

Downfall

Newton's zeroth law was accepted by scientists for more than two centuries, and not just because it fit in with some philosophical or theological intuitions. It was accepted because it worked. Together with Newton's other laws of motion and his law of gravitation, the zeroth law serves to define the mathematical discipline—classical mechanics—that accounts with wonderful precision for the motion of the planets and their moons, the bewildering behavior of gyroscopes, and many other phenomena. And it works brilliantly in chemistry, too.

But it doesn't always work. In fact, the conservation of mass can fail quite spectacularly. At the Large Electron-Positron Collider (LEP), which operated at the CERN laboratory near Geneva through the 1990s, electrons and positrons (antielectrons) were accelerated to velocities within about one part in a hundred billionth (10^{-11}) of the speed of light. Speeding around in opposite directions, the particles smashed into each other, producing a lot of debris. A typical collision might produce ten π mesons, a proton, and an antiproton. Now let's compare the total masses, before and after:

$$\text{electron} + \text{positron:} \quad 2 \times 10^{-28} \text{ gram}$$
$$\text{10 pions} + \text{proton} + \text{antiproton:} \quad 6 \times 10^{-24} \text{ gram}$$

What comes out weighs about *thirty thousand times* as much as what went in. Oops.

Few laws have ever appeared more fundamental, more successful, and more carefully verified than the conservation of mass. Yet here it's gone completely awry. It's as if a magician dropped two peas into her hat and pulled out a few dozen rabbits. But Mother Nature is no cheap trickster; her "magic" is deep truth. We've got some explaining to do.

Does Mass Have an Origin?

As long as mass was thought to be conserved, there was no sense in asking what its origin is. It's always the same. You might as well ask what the origin of 42 is. (Actually there is an answer of sorts. If mass is conserved except when God manufactures elementary particles, then God is the origin of mass. That was Newton's answer. But it's not the kind of explanation we'll be pursuing in this book.)

In the framework of classical mechanics, no answer to the question "What is the origin of mass?" could possibly make sense. Trying to build massive objects from massless ones leads to contradictions. There are many ways to see this. For example:

- The soul of classical mechanics is the equation $F = ma$. This equation relates the dynamical concept of force (F), which summarizes the pull and tug felt by a body, to the kinematic concept of acceleration (a), which summarizes how the body moves in response. The mass (m) mediates between those two concepts. In response to a given force, a body with a small mass will pick up speed faster than a body with a large mass. A body of zero mass would go crazy! In order to figure out how it should move, it would have to divide by zero, which is a no-no. So bodies had better have mass to begin with.
- According to Newton's law of gravitation, each body exerts gravitational influence that is proportional to its mass. In trying to imagine that a body with nonzero mass can be assembled from building blocks without mass, you run smack into a contradiction. The gravitational influence of each building block is zero, and no matter how many times you add zero influence to zero influence you still get zero influence.

But if mass is not conserved—and it's not!—we can seek its origin. It's not bedrock. We can dig deeper.

3

Einstein's Second Law

Einstein's "second law," $m = E/c^2$, raises the question whether mass can be understood more deeply as energy. Can we build, as Wheeler put it, "Mass Without Mass"?

WHEN I WAS ABOUT TO BEGIN TEACHING at Princeton, my friend and mentor Sam Treiman called me into his office. He had some wisdom to share. Sam pulled a well-worn paperback manual from his desk and told me, "During World War II the Navy had to train recruits to set up and operate radio communications in a hurry. Many of those recruits were right off the farm, so bringing them up to speed was a big challenge. With the help of this great book, the Navy succeeded. It's a masterpiece of pedagogy. Especially the first chapter. Take a look."

He handed me the book, opened to the first chapter. That chapter was titled "Ohm's Three Laws." I was familiar with one Ohm's law, the famous relation $V = IR$ that connects voltage (V), current (I), and resistance (R) in an electric circuit. That turned out to be Ohm's first law.

I was very curious to find out what Ohm's other two laws were. Turning the fragile, yellowed pages, I soon discovered that Ohm's second law is $I = V/R$. I conjectured that Ohm's third law might be $R = V/I$, which turned out to be correct.

Finding New Laws, Made Simple

Now for anyone who's had experience with elementary algebra, it's so immediately obvious that those three laws are all equivalent to each other that this story becomes a joke. But there's a deep point to it. (There's also a shallow point, which I think is the one Sam wanted me to absorb. When teaching beginners, you should try to say the same thing several times in slightly different ways. Connections that are obvious to a pro might not come automatically to the beginner. And those students who see you belaboring the obvious won't mind. Very few people get offended when you make them feel clever.)

The deep point connects with a statement made by the great theoretical physicist Paul Dirac. When asked how he discovered new laws of nature, Dirac responded, "I play with equations." The deep point is that different ways of writing the same equation can suggest very different things, even if they are logically equivalent.

Einstein's Second Law

Einstein's second law is

$$m = E/c^2$$

Einstein's first law is of course $E = mc^2$. Famously, that first law suggests the possibility of getting large amounts of energy from small amounts of mass. It calls to mind nuclear reactors and nuclear bombs.

Einstein's second law suggests something quite different. It suggests the possibility of explaining how mass arises from energy. "Second law" is a misnomer, actually. In Einstein's original 1905 paper, you do not find the equation $E = mc^2$. What you find is $m = E/c^2$. (So maybe we should call that Einstein's zeroth law.) In fact, the title of that paper is a question: "Does the Inertia of a Body Depend on Its Energy Content?" In other words: can some of a body's mass arise from the energy of the stuff it contains? Right from the start Einstein was thinking about the conceptual foundations of physics, not about the possibility of making bombs or reactors.

The concept of energy is much more central to modern physics than the concept of mass. This shows up in many ways. It is energy, not mass, that is truly conserved. It is energy that appears in our fundamental equations, such as Boltzmann's equation for statistical mechanics, Schrödinger's equation for quantum mechanics, and Einstein's equation for gravity. Mass appears in a more technical way, as a label for irreducible representations of the Poincaré group. (I won't even try to explain that statement—fortunately, just the act of stating it conveys the point.)

Einstein's question, therefore, lays down a challenge. If we can explain mass in terms of energy, we'll be improving our description of the world. We'll need fewer ingredients in our world-recipe.

With Einstein's second law, it becomes possible to think of a good answer to the question we earlier debunked. What is the origin of mass? It could be energy. In fact, as we'll see, it mostly is.

FAQ

Here are two excellent questions that people frequently ask me when I give public lectures about the origin of mass. If they occurred to you, congratulations! These questions raise basic issues about the possibility of explaining mass in terms of energy.

Question 1: If $E = mc^2$, then mass is proportional to energy. So if energy is conserved, doesn't that mean that mass will be conserved, too?

Answer 1: The short answer is that $E = mc^2$ really applies only to isolated bodies at rest. It's a pity that this equation, the equation of physics that is best known to the general public, is actually a little cheesy. In general, when you have moving bodies, or interacting bodies, energy and mass aren't proportional. $E = mc^2$ simply doesn't apply.

For a more detailed answer, take a look at Appendix A: "Particles have Mass, the World has Energy."

Question 2: How can something made from massless building blocks feel gravitational forces? Didn't Newton tell us that the gravitational force a body feels is proportional to its mass?

Answer 2: In his law of gravitation, Newton indeed told us that the gravitational force felt by a body is proportional to its mass. But Einstein, in his more accurate theory of gravity, general relativity, tells us something different. The complete story is quite complicated to describe, and I won't try to do it here. Very roughly speaking, what happens is that where Newton would say the force is proportional to m, Einstein's more accurate theory says it's proportional to E/c^2. As we discussed in the previous question and answer, those aren't the same thing. They are almost equal for isolated, slowly moving bodies, but they can be very different for interacting systems of bodies, or for bodies moving at close to the speed of light.

In fact, light itself is the most dramatic example. The particles of light, photons, have zero mass. Nevertheless light is deflected by gravity, because photons have nonzero energy, and gravity pulls energy. Indeed, one of the most famous tests of general relativity involves the bending of light by the Sun. In that situation, the gravity of the Sun is deflecting massless photons.

Carrying that thought a step further, one of the most dramatic consequences of general relativity is that you can imagine an object whose gravity is so powerful that it bends photons so drastically as to turn them completely around, even if they're moving straight out at the start. Such an object traps photons. No light can escape it. It is a black hole.

4

What Matters for Matter

What is the world made of? We'll be explaining the origin of matter's mass from pure energy, with 95% accuracy. To attain that sort of precision, we'll have to be very clear about what it is we're talking about. Here we'll be specific about what normal matter is, and what it isn't.

"NORMAL" MATTER IS THE STUFF WE STUDY in chemistry, biology, and geology. It is the stuff we use to build things, and it's what we're made of. Normal matter is also the stuff that astronomers see in their telescopes. Planets, stars, and nebulae are made from the same kind of stuff that we find and study here on Earth. That's the greatest discovery in astronomy.

Recently, however, astronomers have made another great discovery. Ironically, the new great discovery is that normal matter is *not* all there is in the Universe. Not by a long shot. In fact, most of the mass in the Universe as a whole is in at least two other forms: so-called dark matter and dark energy. The "dark" stuff is actually perfectly transparent, which is why it managed to escape notice for hundreds of years. So far it's been detected only indirectly, through its gravitational influence on normal matter (that is, stars and galaxies). We'll have much more to say about the dark side in later chapters.

If you just count up mass, then normal matter is a minor impurity, contributing only 4–5% of the total. But it's where the vast bulk of the structure, information, and love in the world reside.

So I hope you'll agree it's an especially interesting part. And it's the part we understand best, by far.

In the next few chapters we'll account for the origin of 95% of the mass of normal matter, starting from massless building blocks. To make good on that promise, we'll have to be quite specific about what it is we're explaining. (After all, we're quoting numbers.)

Building Blocks

Speculation that matter[1] could be analyzed down to a few types of elementary building blocks goes back at least to the ancient Greeks, but solid scientific understanding came only in the twentieth century. Matter, people ordinarily say, is made of atoms. The great physicist Richard Feynman, near the beginning of his famous *Feynman Lectures on Physics*, made a big point of it:

> If, in some cataclysm, all of scientific knowledge were to be destroyed, and only one sentence passed on to the next generations of creatures, what statement would contain the most information in the fewest words? I believe it is the *atomic hypothesis* (or atomic *fact*, or whatever you wish to call it) that *all things are made of atoms* . . . (italics in original).

The great and most useful "fact" that all things are made of atoms is, however, incomplete in three important ways. (Like Newton's zeroth law, or the greatest discovery in astronomy, it's a profound truth in Bohr's sense—that is, it's also *profoundly* false.)

One aspect of its incompleteness is the existence of dark matter and dark energy, as we've already mentioned. Their existence was barely suspected in 1963, when Feynman's lectures were published. A few astronomers, starting with Fritz Zwicky, were on to what they called a missing mass problem as early as 1933. But the anomalies

1. Starting here, and until Chapter 8, I will drop the adjective *normal* in talking about normal matter, and just say *matter*. We won't revisit the dark side until then.

they noticed were just a few among many in the embryonic science of observational cosmology, and few people took them seriously until much later. In any case, the existence of dark matter and dark energy doesn't really affect Feynman's point. In the initial steps of reconstructing science after a cataclysm, awareness of dark matter and dark energy would be a burdensome luxury.

Two other, much more down-to-earth refinements really are central. They really ought to be included even in our one-sentence message to the next generations of creatures, even though it might lead to the one sentence becoming a longish run-on sentence of the kind that my teachers taught me to avoid and that would cost you points in an SAT essay even though Henry James and Marcel Proust got very famous despite using just those kind of sentences because it's okay if you're writing literature but not okay if you're just conveying information.

First, there's the matter of light. Light is a most important element of "all things," and of course it is quite distinct from atoms. There is a natural instinct to regard light as something quite different from matter, as immaterial or even spiritual. Light certainly *appears* to be quite different from tangible matter—the kind that hurts your toe when you kick it, or whose streams and winds can push you around. It would be appropriate to tell Feynman's hypothetical post-cataclysmians that light is another form of matter they could understand. You might even tell them that light too is made of particles—particles known as photons.

Second, atoms are not the end of the story. They are made of more fundamental building blocks. A few hints along those lines would jump-start the post-cataclysmians on their way toward scientific chemistry and electronics.

The relevant facts can be summarized in a few sentences. (I won't try to do it in one.) All things are made from atoms and photons. Atoms in turn are made from electrons and atomic nuclei. The nuclei are very much smaller than the atoms as a whole (they have roughly one-hundred-thousandth, or 10^{-5}, the radius), but they contain all the positive electric charge and nearly all the mass of the atom—more than 99.9%. Atoms are held together by electrical attraction between the electrons and the nuclei. Finally, nuclei in turn are made from protons and neutrons. The nuclei are held together by another force, a force that is much more powerful than the electric force but acts only over short distances.

That account of matter reflects the state of knowledge as of 1935 or so. It is what you still find in most introductory physics textbooks. To do justice to our best modern understanding, we'll need to qualify, modify, and refine almost every word of it. For example, we've come to understand that protons and neutrons themselves are complicated objects, made from more elementary quarks and gluons. We'll get to the refinements in later chapters. But the 1935 picture is useful as a convenient sketch, a rough outline that's good enough to let us see clearly just what it is we need to do, if we're going to understand the origin of mass.

Most mass is in atomic nuclei, and nuclei are made from protons and neutrons. Electrons contribute much less than 1%, and photons even less. So the problem of the origin of mass, for normal matter, takes very definite shape. To find the origin of most of the mass of matter—more than 99%—we must find the origin of mass of protons and neutrons, and must understand how those particles combine to make atomic nuclei. No more, no less.

5

The Hydra Within

The program to understand atomic nuclei "the old-fashioned way," as systems of protons and neutrons stuck together or orbiting one another, self-destructed. Physicists searching for forces between persistent particles instead uncovered a bewildering new world of transformation and instability.

In 1930, the direction for the next step along the path toward a complete theory of matter was clear. The inward journey of analysis had reached the core of atoms, their nuclei.

Most of the mass of matter, by far, is locked up in atomic nuclei. And the electric charge concentrated in atomic nuclei sets up electric fields that control the motion of the surrounding electrons. The nuclei, because they are much heavier, ordinarily move much more slowly than the electrons. Electrons are the actors in chemical and biological processes (not to mention electronics), but the nuclei lurk behind the scenes, writing the script.

Although in biology, chemistry, and electronics atomic nuclei mostly stay backstage, they star in the story of stars. It is from nuclear rearrangement and transformation that stars, including of course our own Sun, derive their energy. So the importance of understanding atomic nuclei was, and is, obvious.

But in 1930 that understanding was primitive, and the challenge to improve it rose to the top of the physics agenda. In his lectures Enrico Fermi would draw a hazy cloud at the center of his diagram of an atom, with the label "Here be Dragons," as in ancient maps of unexplored regions. Here was the wild frontier that had to be explored.

Fermi's Dragons

It was clear from the start that essentially new forces rule the nuclear world. The classic forces of pre-nuclear physics are gravity and electromagnetism. But the electric forces acting in nuclei are repulsive: the nuclei have overall positive charge, and like charges repel. Gravitational forces, acting on the tiny amount of mass in any one nucleus, are far too feeble to overcome the electric repulsion. (We'll have a lot more to say about the feebleness of gravity in the second part of this book.) A new force was required. It was dubbed the strong force. In order to keep nuclei so tightly bound together, the strong force had to be more powerful than any previously known force.

It took decades of experimental effort and theoretical ingenuity to discover the fundamental equations that govern what goes on in atomic nuclei. What's amazing is that people were able to find them at all.

The obvious difficulty is simply that it is hard to observe those equations at work, because atomic nuclei are very small. They are roughly a hundred thousand times smaller even than atoms. This takes us a million times beyond nanotechnology. Nuclei are in the domain of micro-nanotechnology. In trying to manipulate nuclei with macroscopic tools—say, to set the scale, an ordinary tweezer—we're worse off than some giant trying to pick up a grain of sand using a pair of Eiffel towers for chopsticks. It's a tough job. To explore the nuclear domain, wholly new experimental techniques had to be invented, and exotic kinds of instruments constructed. We'll visit an ultrastroboscopic nanomicroscope (known as the Stanford Linear Accelerator, SLAC) and a creative destruction powerhouse (known as the Large Electron-Positron Collider,

LEP), where discoveries central to our story took place, in the next chapter.

Another difficulty is that the micronanocosm turns out to follow new rules unlike anything that came before. Before they could do justice to the strong interaction, physicists had to discard ways of thinking that come naturally to humans, and replace them with strange new ideas. We'll discuss those ideas in depth over the next few chapters. They are so strange, that if I just asserted them as facts they might not—they should not—seem credible.[1] Some of the new ideas are completely unlike anything that came before. They might appear to contradict—and might actually contradict!—things you learned in school. (It depends on what school you went to, and when.) This short chapter is meant to indicate why we were driven to revolution. It serves to connect the traditional account of nuclear physics (the account you still find in most of the high school and introductory college physics textbooks I've looked at) with the new understanding.

Wrestling with Dragons

James Chadwick's discovery of the neutron in 1932 was a landmark. After Chadwick's discovery, the path to understanding seemed straightforward. It seemed that the building blocks of nuclei had been discovered. They are protons and neutrons, two kinds of particles that weigh about the same (the neutron is about 0.2% heavier) and have similar *strong* interactions. The most obvious differences between protons and neutrons are that the proton has positive electric charge, whereas the neutron is electrically neutral; and that a neutron in isolation is unstable, decaying, with a lifetime of about fifteen minutes into a proton (plus an electron

1. Toward the end of this book, I'll discuss other strange ideas, for which the evidence is as yet much less convincing. I want you to appreciate the difference!

and an antineutrino). Simply by adding together protons and neutrons, you could make model nuclei with different charges and masses that roughly matched those of known nuclei.

To understand and sharpen that modeling, it seemed, was just a matter of measuring the forces that act among protons and neutrons. Those forces would hold the nuclei together. The equations describing those forces would be the theory of the strong interaction. By solving that theory's equations, we could both check the theory and make predictions. Thus we would write a neat new chapter of physics called "Nuclear Physics" whose centerpiece would be a nice "nuclear force" described by a simple, elegant equation.

Inspired by that program, experimenters studied close encounters of protons with other protons (or neutrons, or other nuclei). We call this kind of experiment, where you shoot one kind of particle at another and study what comes out, a scattering experiment. The idea is that by studying how the protons and neutrons swerve, or (as we say) scatter, you can reconstruct the force that's responsible.

This straightforward strategy failed miserably. First, the force got very complicated. It was found to depend not only on the distance between particles but also on their velocities and on the directions of their spins[2] in a complicated, mixed-up way. It quickly became clear that no simple and beautiful law for the force, worthy of standing beside Newton's law of gravity or Coulomb's law of electricity, was going to emerge.

Second, and even worse: the "force" wasn't a force. What you find when you shoot two energetic protons together is *not* simply that they swerve. Instead, often more than two particles come out, and they are not necessarily protons. Indeed, many new kinds of

---

2. Both protons and neutrons are always spinning: we say they have an intrinsic, fundamental spin. We'll have much more to say about the property of fundamental spin later. It plays a crucial role in modern ideas about the ultimate unification of forces.

particles were discovered in this way, as physicists did scattering experiments at high energy. The new particles—eventually dozens were discovered—are unstable, so we do not ordinarily encounter them in nature. But when they were studied in detail it seemed that their other properties—in particular, their strong interactions and their size—are similar to those of protons and neutrons.

After these discoveries, it became unnatural to consider protons and neutrons by themselves, or to think that the fundamental issue was to determine the forces between them. Instead "nuclear physics," as traditionally conceived, became part of a bigger subject, involving all the new particles and the apparently complicated processes whereby they are created and decay. A new name was invented to describe the new zoo of elementary particles, this new genus of dragons. They are called hadrons.[3]

Hydra

Experience in chemistry suggested a possible explanation for all this complexity. Maybe protons and neutrons and the other hadrons are not elementary particles. Maybe they are made from more basic objects with simpler properties.

Indeed, if you do the same sorts of experiments for atoms and molecules that were done with protons and neutrons, studying what comes out of their close encounters, you'll also find complicated results. You could have rearrangements and break-ups to form new kinds of molecules (or excited atoms, ions, or radicals)—in other words, chemical reactions. It is only the underlying electrons and nuclei that obey a simple force law. The atoms and molecules made from many electrons and nuclei do not. Could there be a similar story for protons, neutrons, and their newly discovered relatives? Could their apparent complexity arise out of

3. *Not* a typo.

their intricate construction from more basic building blocks that obey radically simpler laws?

Breaking something to bits might be crude, but you might think it's a foolproof way to find out what that something is made out of. If you bang atoms together hard enough, they break up into their constituent electrons and nuclei. Thus are their underlying building blocks revealed.

But the search for simpler building blocks inside protons and neutrons ran into a bizarre difficulty. If you bang protons together really hard, what you find coming out is . . . more protons, sometimes accompanied by their hadronic relatives. A typical outcome would be, you collide two protons at high energy, and out come three protons, an antineutron, and several π mesons. The total mass of the particles that come out is more than what went in. We discussed that possibility earlier, and it's back again to haunt us. Instead of discovering smaller, lighter building blocks by going to higher and higher energies, and making more violent collisions, you just find more of the same. Things don't appear to get simpler. It's as if you smashed together two Granny Smith apples, and got three Granny Smiths, a Red Delicious, a cantaloupe, a dozen cherries, and a pair of zucchini!

Fermi's dragon had become the nightmarish Hydra of myth. Chop Hydra up, and more Hydras spring to life from the pieces.

There *are* simpler building blocks. But their fundamental "simplicity" includes some weird and paradoxical behavior that makes them both theoretically revolutionary and experimentally elusive. To understand them—even to *perceive* them—we need to make a fresh start.

6

The Bits Within the Its

Introduced in a theoretical improvisation and never observed in isolation, quarks at first seemed a convenient fiction. But when they showed up in ultrastroboscopic nanomicroscopic snapshots of protons, quarks became an inconvenient reality. Their strange behavior called basic principles of quantum mechanics and relativity into question. A new theory reinvented quarks as ideal objects of mathematical perfection. The equations of the new theory also demanded new particles, the color gluons. Within a few years, people were taking pictures of both quarks and gluons, at powerhouses of creative destruction built for the purpose.

THE TITLE OF THIS CHAPTER has two meanings. The first is simply that there are littler bits within what not so long ago were thought to be basic building blocks of ordinary matter, protons and neutrons. These littler bits are called quarks and gluons. Of course, knowing something's name does not tell you what it is, as Shakespeare had Romeo explain:

> What's in a name? That which we call a rose
> By any other name would smell as sweet.

Which brings us to the second, more profound meaning. If quarks and gluons were just another layer in a never-ending onion of complex structure within structure, their names would provide impressive-sounding buzzwords you could show off at a cocktail party, but they themselves would be of interest only to specialists. Quarks and gluons, are, however, not "just another layer." When properly understood, they change our understanding of the nature of physical reality in a fundamental way. For quarks and gluons are bits in another and much deeper sense, the sense we use when we speak of bits of information. To an extent that is qualitatively new in science, they are *embodied ideas.*

For example, the equations that describe gluons were discovered before the gluons themselves. They belong to a class of equations invented by Chen Ning Yang and Robert Mills in 1954 as a natural mathematical generalization of Maxwell's equations of electrodynamics. The Maxwell equations have long been renowned for their symmetry and power. Heinrich Hertz, the German physicist who proved experimentally the existence of the new electromagnetic waves Maxwell had predicted (what we now call radio waves), said of Maxwell's equations:

> One cannot escape the feeling that these mathematical formulae have an independent existence and an intelligence of their own, that they are wiser than we are, wiser even than their discoverers, that we get more out of them than was originally put into them.

The Yang-Mills equations are like Maxwell's equations on steroids. They support many kinds of charges, instead of just the one kind (electric charge) that appears in Maxwell's equations, and they support symmetry among those charges. The specific version that applies to the real-world gluons of the strong interaction, which uses three charges, was proposed by David Gross and me in 1973. The three kinds of charges that appear in the theory of the strong interaction are usually called color charges, or simply color, although of course they have nothing to do with color in the usual sense.

We'll discuss the nuts and bolts of quarks and gluons much more below. The point I want to emphasize here, from the beginning, starting with the title, is that quarks and gluons, or more precisely their fields, are mathematically complete and perfect objects. You can describe their properties completely using concepts alone, without having to supply samples or make any measurements. And you can't change those properties. You can't fiddle with the equations without making them worse (indeed, inconsistent). Gluons are the objects that obey the equations of gluons. The its *are* the bits.

But enough of this all-too-free rhapsody! Pure mathematics is chockablock with beautiful ideas. The special music of physics lies in harmony between beautiful ideas and reality. It's time to bring in some reality.

Quarks: Beta Release

By the early 1960s, experimenters had discovered dozens of hadrons, with different masses, lifetimes, and intrinsic rotation (spin). The orgy of discovery soon led to a hangover, as the mere accumulation of curious facts, absent any deeper meaning, became mind-numbing. In his 1955 Nobel Prize acceptance speech, Willis Lamb joked

> When the Nobel Prizes were first awarded in 1901, physicists knew something of just two objects which are now called "elementary particles": the electron and the proton. A deluge of other "elementary" particles appeared after 1930; neutron, neutrino, μ meson, π meson, heavier mesons, and various hyperons. I have heard it said that "the finder of a new elementary particle used to be rewarded by a Nobel Prize, but such a discovery now ought to be punished by a $10,000 fine."

In this situation, Murray Gell-Mann and George Zweig made a great advance in the theory of the strong interaction by proposing the quark model. They showed that patterns among the masses, lifetimes, and spins of hadrons click into place if you imagine that hadrons are assembled from a few more basic kinds of objects, which Gell-Mann named quarks. Dozens of hadrons could be understood, at least roughly, as different ways of putting together just three varieties, or *flavors*, of quarks: up *u*, down *d*, and strange *s*.[1]

How do you build dozens of hadrons from a few flavors of quarks? What are the simple rules behind the complicated patterns?

The original rules were improvised to fit the observations, and they're a bit peculiar. They defined what is called the quark model. According to the quark model, there are just two basic body plans for hadrons. *Mesons* are constructed from a quark and an antiquark. *Baryons* are constructed from three quarks. (There are also antibaryons, constructed from three antiquarks.) Thus there are just a handful of possibilities for combining quarks and antiquarks of different flavor to make mesons: you can combine *u* with anti-*d* ($\bar{d}$), or *d* with $\bar{s}$, and so forth. Similarly for baryons, there are only a few possible combinations.

According to the quark model, the great diversity of hadrons comes not so much from which pieces you put together as from how you assemble them. Specifically, a given set of quarks can be arranged in different spatial orbits, with their spins aligned in different ways, in roughly the same way that pairs or triples of stars can be bound together by gravity.

There's a crucial difference between the submicroscopic "star systems" of quarks and their macroscopic brethren. Whereas macroscopic solar systems, governed by the laws of classical mechanics, can come in all sizes and shapes, the microscopic version cannot. For microscopic systems, which obey the laws of quantum mechanics, there are restrictions on the allowed orbits and spin

---

1. The flavors of quarks should not be confused with their color charges. Color charge is a different, additional property. There are *u* quarks with a unit of red color charge, *u* quarks with a unit of blue color charge, and so forth. Thus with 3 flavors and 3 colors, we have $3 \times 3 = 9$ kinds altogether.

alignments.[2] We say that the system can be in different quantum *states.* Each allowed configuration of orbits and spins—each state—will have some definite total energy.

(Confession and preview: I'm being a little sloppy here, so as not to pile on too many complications at once. According to modern quantum mechanics, the correct way to describe the state of a particle is in terms of its wave function, which describes the probability of finding it at different places, rather than in terms of an orbit it follows. We'll talk about this more in Chapter 9. The orbit picture is a relic of the so-called old quantum mechanics. It is easy to visualize, but it can't be used for precision work.)

This setup for using quarks to understand hadrons runs strictly parallel to the way we use electrons to understand atoms. The electrons in an atom can have orbits of different shapes and can align their spins in different directions. Thus the atom can be in many different states, with different energies. The study of the possible states is a vast subject known as atomic spectroscopy. We use atomic spectroscopy to reveal what distant stars are made of, to design lasers, and for many other things. Because atomic spectroscopy is so relevant to the quark model, and is extremely important in itself, let's take a moment to discuss it.

A hot gas, such as you find in a flame or a stellar atmosphere, contains atoms in different states. Even atoms with the same kind of nuclei and the same number of electrons can have their electrons in different orbits or with their spins aligned in different ways. These states have different energies. States of high energy can decay into states with lower energy, emitting light. Because energy is conserved overall, the energy of the emitted photon, betrayed by its color, encodes the difference in energy between the initial and final states. Every kind of atom emits light from a

---

2. Strictly speaking, the laws of quantum mechanics are universal: they apply to macroscopic star systems just as well as to microscopic ones like atoms. For macroscopic systems, however, the quantum restrictions on orbits have no practical significance, because the spacing between allowed orbits is minuscule.

characteristic palette of colors. Hydrogen atoms emit one set of colors, helium atoms an entirely different set, and so forth. Physicists and chemists call that palette the *spectrum* of the atom. The spectrum of an atom functions as its signature and can be used to identify it. When you put light through a prism the different colors get separated, and the spectrum literally resembles a barcode.

It's because the spectra we observe in starlight match the spectra we observe in terrestrial flames that we can be confident that distant stars are made of the same basic kind of material as we find on Earth. Also, because the light from distant stars may take billions of years to reach us, we can check whether the laws of physics that operate today are the same as those that operated long ago. So far, the evidence is that they are. (But there are good reasons to think that the *very* early universe, which we can't see directly, at least in ordinary light, was governed by essentially different laws. We'll discuss that later.)

Atomic spectra give us a lot of detailed guidance for contructing models of the internal structure of atoms. To be valid, a model must predict states whose energy differences match the pattern of colors that spectra reveal. Much of modern chemistry takes the form of a dialogue. Nature speaks in spectra; chemists reply in models.

With that background in mind, let's return to the quark model of hadrons. The same ideas come into play, with one major refinement. In atoms, the difference in energy between any two states of the electrons is relatively small, so the effect of that energy difference on the overall mass of the atom is insignificant. A central idea of the quark model is that for quark "atoms" (that is, hadrons), the energy differences between different states are so large that they contribute to the mass in a big way. Turning it around, exploiting Einstein's second law $m = E/c^2$, we can interpret hadrons with different masses as systems of quarks in different orbital patterns—different quantum states—that have different energies. In other words, we *see* atomic spectra, but we *weigh* hadron spectra. Thus what appeared to be unrelated particles now appear as merely different patterns of motion within a given "atom" of quarks. Using that idea, Gell-Mann and Zweig showed that one could interpret many different observed hadrons as different states of a few underlying quark "atoms."

So far, so easy. Except for the refinement introduced by Einstein's second law, the quark model of hadrons looks like a replay of chemistry. But the devil is in the details, and to see reality in the quark model, one had to turn a blind eye toward some truly fiendish deviltry.

The most wicked assumption is the one we already mentioned, that only meson (quark-antiquark) and baryon (three-quark) body plans are allowed. That assumption includes, in particular, the idea that quarks don't exist as individual particles! For some reason, you had to suppose, the simplest body plan is impossible. Not just inefficient or unstable, but impossible. Nobody wanted to believe that, of course, so people worked hard to smash protons, trying to find particles they could identify as single quarks. They scrutinized the debris minutely. Nobel Prizes and everlasting glory surely would shower down like sweet rain upon the heads of the discoverers. But alas, that Holy Grail proved elusive. No particle that has the properties of a single quark has ever been observed. Eventually this failure to find individual quarks, like the failure of inventors to produce a perpetual motion machine, was elevated to a principle: the Principle of Confinement. But calling it a principle didn't make it less crazy.

More deviltry came in when physicists tried to make fleshed-out models of the internal structure of mesons and baryons using quarks, to account for their masses in detail. In the most successful models, it appeared that when quarks (or antiquarks) are close together, they hardly notice one another. That feeble interaction between quarks was hard to reconcile with the finding that if you tried to isolate one quark—or two—you found you could not. If quarks don't care about one another when they're close, why do they object to being separated when they're far away?

A fundamental force that *grows* with distance would be unprecedented. And it would pose an embarrassing question. If forces between quarks can grow with distance, why is astrology wrong?

After all, the other planets contain lots of quarks. Maybe they could exert a big influence. . . . Well, maybe, but for centuries scientists and engineers have been very successful at predicting the results of delicate experiments and building bridges and designing microchips by ignoring any possible influence of distant objects. Astrology should be made of sterner stuff.

Because a good scientific theory has to explain why astrology is so lame, it had better not contain forces that grow with distance. The old saw "Absence makes the heart grow fonder" may or may not apply to romance, but it's surely a bizarre way for particles to behave.

In software development, often a beta test version is supplied for use by brave early adopters. The beta test version works, more or less, but comes with no guarantees. There will be bugs and missing features. Even the parts that work won't be smoothly polished.

The original quark model was a beta test physical theory. It used peculiar rules. It left basic questions, like why (or whether) quarks could ever be produced in isolation, unanswered. Worst of all, the quark model was vague. It did not come with precise equations for the forces between quarks. In that respect it resembled pre-Newtonian models of the solar system, or pre-Schrödinger (for experts: even pre-Bohr) models of atoms. Many physicists, including Gell-Mann himself, thought that quarks might turn out to be a useful fiction, like the epicycles of the old astronomy or the orbitals of old quantum theory. Quarks, it seemed, might turn out to be useful stopgaps in the mathematical description of nature, not to be taken too literally as elements of reality.

Quarks 1.0: Through an Ultrastroboscopic Nanomicroscope

The theoretical peculiarities of quarks ripened into juicy paradoxes in the early 1970s, when Jerome Friedman, Henry Kendall, Richard Taylor, and their collaborators at the Stanford Linear Accelerator (SLAC) studied protons in a new way.

Instead of bashing protons together and scrutinizing the debris, they photographed proton interiors. I don't want to make that sound easy, because it isn't. To look inside protons, you must use "light" of very short wavelength. Otherwise you'd be trying, in effect, to locate fish by looking for their effect on long ocean waves. The photons for this job are not particles of ordinary light. They lie beyond ultraviolet or even x-rays. A nanomicroscope fit for studying structures a billion times beyond the ken of ordinary optical microscopes requires *extreme* γ-rays.

Also, things move quickly inside protons, so to avoid blurring the picture we must have good time-resolution. Our photons, in other words, must also be extremely short-lived. We need flashes, or sparks, not long exposures. We're talking about "flashes" that last 10^{-24} second, or less. The photons we need are so short-lived that they themselves can't be observed. That's why they're called *virtual* photons. An ultrastroboscope to look at features that last for a trillionth of a trillionth of the blink of an eye (actually, even less) requires extremely virtual photons. So the "picture" can't be made using the transient "light" that provides the illumination! We have to be cleverer, and work indirectly.

At SLAC people actually shot electrons at protons, and observed electrons emerging after they collided. The emerging electrons have less energy and momentum than when they started. Because energy and momentum are conserved overall, what was lost by the electron had to be carried away by the virtual photon, and transmitted to the proton. This often causes the proton to break apart in complicated ways, as we've discussed. The stroke of genius—the new approach that won Friedman, Kendall, and Taylor their Nobel Prize—was to ignore all those complications and just keep track of the electron. In other words, we just go with the flow (of energy and momentum).

In this way, by accounting for the flow of energy and momentum, we can figure out what kind of virtual photon was involved, event by event, even though we don't "see" that photon directly. The energy and momentum of the virtual photon are precisely the energy and momentum *lost* by the electron. By measuring the probability that different kinds of virtual photons, with different energies and momenta (corresponding to different lifetimes and wavelengths), "encountered something" and got absorbed, we can piece together a snapshot of the proton's interior. The procedure is similar in spirit to the way we reconstruct a picture of a human body's interior by measuring how x-rays get absorbed, although the details are considerably more complicated. Suffice it to say that some very fancy image processing is involved.

Now of course the interiors of protons don't really look like anything you've ever seen, or could see. Our eyes were not designed (ahem, did not evolve) to resolve such small distances and times, so any visual representation of the ultrastrobonanomicroworld must be a mixture of caricature, metaphor, and cheat. With that warning, please look now at the panels of Figure 6.1. We'll be discussing various aspects of them further below.

In presenting these pictures, I've used a trick I owe to Richard Feynman. As we've noted, things move fast inside a proton. To slow things down, we imagine that the proton is moving past us at very nearly the speed of light. (In Chapter 9, we'll discuss how protons look if we don't use Feynman's trick.) From the exterior, the proton then comes to look like a pancake, flattened in the direction of motion. This is the famous Fitzgerald-Lorentz contraction of special relativity. More important for our purposes is another famous relativistic effect, time dilation. Time dilation means that time appears to flow more slowly within a fast-moving object. Thus the stuff inside the protons appears nearly frozen in place. (It shares the overall motion of the proton as a whole, of course.) Fitzgerald-Lorentz contraction and time dilation have been explained in hundreds of popular books on relativity, so rather than belaboring them here, I'll just use them.

It's important to emphasize that quantum mechanics is absolutely essential for describing even the most elementary observations about proton interiors. In particular, the indeterminism for which quantum mechanics is famous, and which caused Einstein

Figure 6.1 Pictures of the interior of a proton. a. A proton moving at nearly the speed of light appears flattened in the direction of motion, according to the theory of relativity. b. A good guess, before actual snapshots were available, for what the interior might look like. The reasoning behind this (wrong) guess is explained in the text. c–d. Two actual snapshots. Because quantum-mechanical uncertainty is a dominant effect in this domain, each snapshot looks different! Inside are quarks and gluons, also moving at nearly the speed of light. They share the total momentum of the proton, and the sizes of the arrows indicate their relative shares. e–f. If you look with finer resolution, more details appear. For example, you may find that what appeared to be a quark resolves into a quark and a gluon, or that a gluon resolves into a quark and an antiquark.

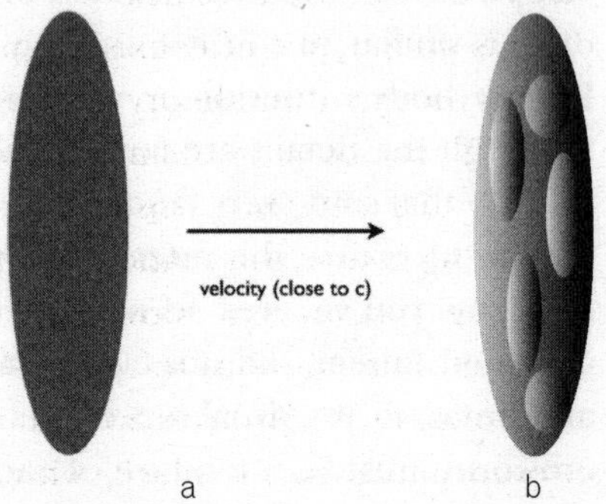

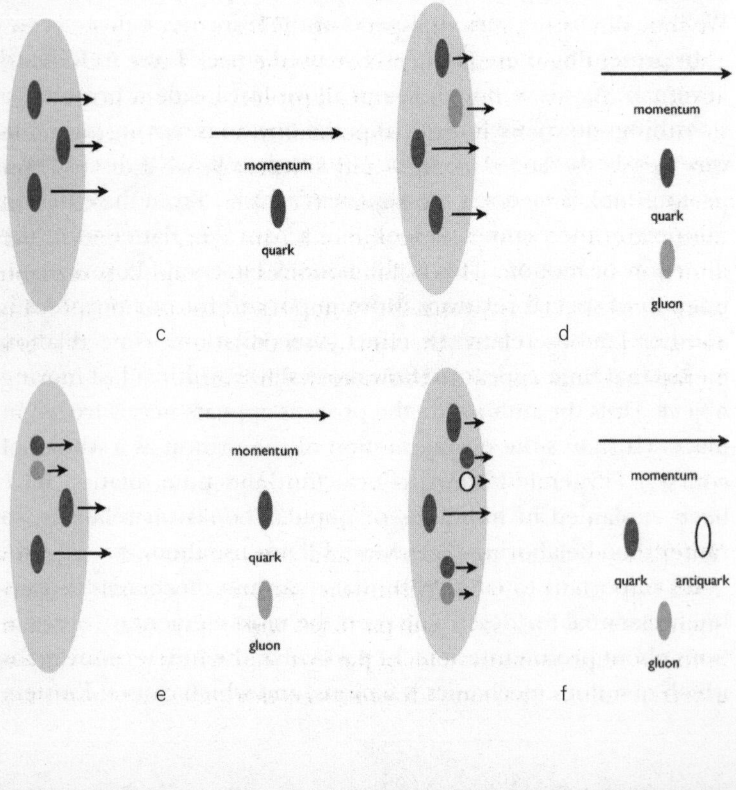

such anguish, hits you in the face. If you take several snapshots of a proton under strictly identical conditions, you see different results. Like it or not, the facts are straightforward and unavoidable. The best we can hope for is to predict the relative probabilities of the different results.

This abundance of coexisting possibilities in the phenomena, and in the quantum theory that describes them, defies traditional logic. The success of quantum theory in describing reality transcends and in a sense unseats classical logic, which depends on one thing being "true," and its contraries "false." But this is a creative destruction that allows new imaginative constructions. For example, it enables us to reconcile two seemingly contradictory ideas about what protons are. On the one hand, the interior of a proton is a dynamic place, with things changing and moving around. On the other hand, all protons everywhere and everywhen behave in exactly the same way. (That is, each proton gives the same probabilities!) If a proton at one time is not the same as itself at a different time, how can all protons be identical??

Here's how. Every individual possibility *A* for the proton's interior evolves in time into a new and different possibility, say *B*. But meanwhile, some other possibility *C* evolves into *A*. So *A* is still there; the new copy replaces the old. And more generally, even though each individual possibility evolves, the complete distribution of possibilities remains the same. It is like a smoothly flowing river, which always looks the same even though every drop of it is in flux. Wade deeper into this river in Chapter 9.

Partons, Put-Ons, and Putdowns

The pictures taken by Friedman and company presented both a revelation and a puzzle. Within the pictures, you could discern some little entities, little sub-particles, inside protons. Feynman, who was responsible for a lot of the image processing, called these internal entities "partons" (for particles that are *parts* of protons).

That infuriated Murray Gell-Mann. As I learned firsthand, when I first met Gell-Mann. He asked me what I was working on. I made the mistake of saying, "I'm trying to improve the parton model." I've heard that confession is good for the soul, so here I confess that it wasn't entirely in innocence and ignorance that I mentioned partons. I was curious to see how Gell-Mann would react to his rival's idiom. As Ishmael wrote of his first encounter with Captain Ahab, reality outran anticipation.

Gell-Mann gave me a quizzical look. "Partons?" Stage pause, facial expression of deep concentration. "Partons?? What are partons?" Then he paused again and looked very thoughtful, until suddenly his face brightened. "Oh, you must mean those *put-ons* that Dick Feynman talks about! The particles that don't obey quantum field theory. There's no such thing. They're just quarks. You shouldn't let Feynman pollute the language of science with his jokes." Finally, with a quizzical expression, but a tone of authority: "Don't you mean quarks?"

Some of the entities Friedman and company found really did appear to be quarks. They had both the funny fractional electric charges and the precise amount of spin that quarks were supposed to have. But protons also contain bits that don't look like quarks. They were later interpreted as color gluons. So both Gell-Mann and Feynman had valid points: there are quarks inside, but also other things.

Too Simple

At my alma mater, the University of Chicago, they sell sweatshirts that read

> That works in *practice*, but what about in *theory*?

Both Gell-Mann's quarks and Feynman's partons had the annoying feature that they worked well in practice but not in theory.

We've already discussed how the quark model helped to organize the hadron zoo, but only by using crazy rules. The parton model used different crazy rules, this time to interpret those pictures of the proton's interior. The rules of the parton model are very simple: you're supposed to assume, for purposes of calculation, that the bits inside the proton—quarks, partons, whatever you want to call them—have *no* internal structure and *no* interactions with each other. Of course they do interact; otherwise, protons would just fly apart. But the idea of the parton model is that you get a good approximate description of what happens in a very short time, over very short distances, by ignoring the interactions. And it is that short-time, short-distance behavior you access with the SLAC ultrascroboscopic nanomicroscope. So the parton model says you will get a clean view of the interior using that instrument—as in fact you do. And you should see more basic building blocks, if such there be—as in fact you do.

It all sounds very reasonable, almost intuitively obvious. Nothing much can happen in a very short time, in a very small volume. What's crazy about that?

The trouble is that when you get down to really short distances and really short times, quantum mechanics comes into play. When you take quantum mechanics into account, the "reasonable, almost intuitively obvious" expectation that nothing much can happen in a short time in a small volume comes to seem very naïve.

One way to appreciate why, without going too deep into technicalities, is by considering Heisenberg's uncertainty principles. According to the original uncertainty principle, to pin down a position accurately we must live with a large uncertainty in momentum. An addendum to Heisenberg's original uncertainty principle is required by the theory of relativity, which relates space to time and momentum to energy. This additional principle says that to pin down a time accurately we must live with a large uncertainty in energy. Combining the two principles, we discover that to take high-resolution, short-time snapshots, we must let momentum and energy float.

Ironically, the central technique of the Friedman-Kendall-Taylor experiments, as we mentioned, was precisely to concentrate on measuring the energy and momentum. But there's no contradiction. On the contrary, their technique is a wonderful example of Heisenberg's uncertainty principle cleverly harnessed to give certainty. The point is that to get a sharply resolved space-time image you can—and must—combine results from *many* collisions with different amounts of energy and momentum going into the proton. Then, in effect, image processing runs the uncertainty principle backwards. You orchestrate a carefully designed sampling of results at different energies and momenta to extract accurate positions and times. (For experts: you do Fourier transforms.)

Because to get a sharp image you need to allow for a big spread in energy and momentum, you must in particular allow for the possibility of large values. With large values of energy and momentum you can access a lot of "stuff"—for instance, lots of particles and antiparticles. These *virtual* particles come to be and pass away very quickly, without going very far. Remember, we've only run into them in the process of making a short-time, high-resolution snapshot! We don't see them, in any ordinary sense, unless we supply the energy and momentum needed to create them. And even then what we see is not the original, undisturbed virtual particles (the kind that appear and disappear spontaneously) but real particles we can use to recreate the original virtual particles by image processing.

Viruses can come to life only with the help of more complex organisms. Virtual particles are still more insubstantial, for they need external help to come into *existence.* Nevertheless, they appear in our quantum-mechanical equations, and according to those equations the virtual particles affect the behavior of the particles we see.

And so it seemed reasonable to expect that the virtual particles should have big effects when we're dealing with particles that interact strongly, such as the things that make protons. Sophisticated quantum-mechanicians expected that the closer and faster you looked inside protons, the more virtual particles and complexities you'd see. And so the Friedman-Kendall-Taylor approach wasn't considered very promising. The ultrastroboscopic nanomicroscopic snapshot would just be a blur.[3]

But it wasn't a blur. It was those infuriating partons. A famous piece of wise advice from Einstein is to "Make everything as simple as possible, but not simpler." Partons were too simple.

Asymptotic Freedom (Charge Without Charge)

Let's imagine we're virtual particles. We pop into existence and have to decide what to do in our all-too-brief lifetime. (That's not so hard to imagine.) We sniff around. Suppose there's a positively charged particle in the region. If we're negatively charged, we find that particle attractive and try to snuggle up to it. If we're positively charged, we find that other particle repulsive, or at least competitive and possibly intimidating, and we move away. (Neither is that.)

Individual virtual particles come and go, but together they make the entity we call empty space into a dynamic medium. Due to the behavior of virtual particles a (real) positive charge is partially *screened.* That is, the positive charge tends to be surrounded by a cloud of compensating negative charges that find it attractive. From far away we do not feel the full strength of the positive charge, because that strength is partially canceled by the negative cloud.[4] Putting it another way, the effective charge grows as you get closer, and shrinks as you move farther away. For a picture of this situation, see Figure 6.2.

3. Actually, a very few extremely smart quantum-mechanicians, most notably James Bjorken, had even more sophisticated arguments indicating that it might work after all.

4. Thus the force falls off faster than 1 over the distance squared, as you'd have without screening.

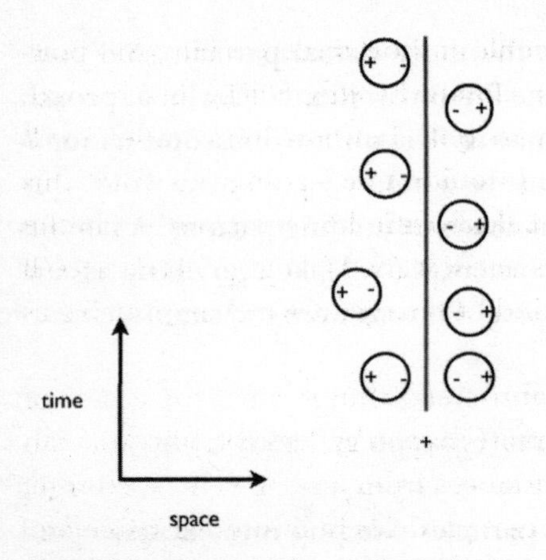

Figure 6.2 The screening of charge by virtual particles. The central world-line shows a positively charged real particle fixed in space—it traces out a vertical line as time advances. That real particle is surrounded by virtual particle-antiparticle pairs, which at random times pop up, briefly separate, and wink out. The positive charge of the real particle attracts the negatively charged member of each virtual pair and repels the positive member. Thus the real particle becomes surrounded, and its positive charge is partially screened, by a negatively charged cloud of virtual particles. From far away we see a smaller effective charge, because the negative virtual cloud partially cancels the central positive charge.

Now this is just the *opposite* of the behavior we want from quarks in the quark model, or partons in the parton model. The quarks of the quark model are supposed to interact weakly when they're close together. But if their effective charge is largest in their immediate vicinity, we'll find just the opposite. They will interact most strongly when they're close together, and more weakly when they are far apart and their charges are screened. The partons of the parton model are supposed to look like simple individual particles when you look closely. But if a thick cloud of virtual particles envelops each parton, we'll see those clouds instead.

Clearly, we'd be much closer to describing quarks if we could arrange to get the opposite of screening—clouds that reinforce, rather than cancel, the central charge. With such *antiscreening* we could have forces that are feeble at short distances but grow powerful far away, thanks to the clouds. Electric charge is screened, not antiscreened, so we have to look elsewhere for a model. We'll find it, of course; otherwise I wouldn't be leading you down this garden path. Just so we can talk about it, let's temporarily call the hypothetical thing that gets antiscreened "churge." (What we'll find is that a generalized kind of charge, color charge, behaves like churge.)

If virtual particle clouds antiscreen churge, then the power of the real, central churge increases as you go farther away. You can get strong forces at large distances from a small central churge, because its entourage of virtual particles builds up its influence. Thus if quarks have churge instead of (or in addition to) electric charge, you can have quarks that interact feebly when close together, as the quark model wants, but powerfully when far apart. You can even do this while avoiding astrology, as I'll explain in a moment. And you can have partons that aren't hidden in thick clouds, because their cloud-inducing power—their effective churge—wanes in their immediate proximity.

What about that unlimited growth of strength with distance, which threatened to bring back astrology? That growth is the result we get for an isolated churged particle. But the big cloud comes at a price. (You might say that expansive clouds are expensive.) It costs energy to create such a disturbance, and it would take an infinite amount of energy to keep it going out to infinite distances. Because the available energy is finite, Nature won't let us create an isolated churged particle. On the other hand, we *can* get away with a system of churged particles whose churges cancel—for example, and most simply, a churged particle and its antiparticle. Virtual particles that are far from both the churge and its canceling antichurge will feel no net attraction, and therefore the clouds won't keep building. All this begins to sound less like justifying astrology, and more like justifying the devilish rules of the quark model! We can both eliminate all the long-range influences and confine whole classes of particles with the same clever idea.

Antiscreening is a horrible word. The standard jargon in physics is *asymptotic freedom*, which may not be much better.[5] The idea is that as the distance to a quark gets closer and closer to zero, the effective color charge deep inside its cloud approaches closer and closer to zero, but never quite gets there. Zero color charge means complete freedom—no influence exerted, and none felt. Such complete freedom is approached, as the mathematicians say, asymptotically.

Whatever you call it, asymptotic freedom is a promising idea for describing quarks and making partons respectable. We'd like to have a theory that includes asymptotic freedom and is also consistent with the basic principles of physics. But is there such a theory?

The rules of quantum mechanics and special relativity are so strict and powerful that it's very hard to build theories that obey both. Those few that do are called relativistic quantum field theories. Because we know only a few basic ways to construct relativistic quantum field theories, it's possible to explore all the possibilities, to see whether any of them leads to asymptotic freedom.

The necessary calculations are not easy to do, but not impossible either.[6] From this work emerged something that every scientist hopes for in a scientific investigation, but rarely finds: a clear, unique answer. Almost all relativistic quantum field theories screen. That intuitive, "reasonable" behavior is, indeed, almost inevitable. But not quite. There is a small class of asymptotically free (antiscreening) theories. They all feature, at their core, the generalized charges introduced by Yang and Mills. Within this small class of asymptotically free theories, there is exactly one that looks even

---

5. When Gross and I discovered asymptotic freedom we were young and naive, and we didn't fully appreciate the importance of naming things in catchy ways. If I had it to do over, I'd call asymptotic freedom something sexy, like "Charge Without Charge." "Asymptotic freedom" was suggested by my good friend Sidney Coleman, whom I forgive.

6. They were much more challenging in 1973 than they are today, because technique has improved.

remotely as if it might describe real-world quarks (and gluons). It's the theory we call quantum chromodynamics, or QCD.

As I've already hinted, QCD is like the quantum version of electrodynamics—quantum electrodynamics, or QED—on steroids. It embodies an enormous amount of symmetry. To do QCD even rough justice, we need to lay some deep foundations using the concept of symmetry. Then we'll build up our description of the theory using drawings and analogies.

The biggest challenge may be to imagine how all those abstractions and metaphors connect with anything real and concrete. To warm up our imaginations, let's start by contemplating a *photograph* of things that don't exist. Behold, in Color Plate 1, a quark, an antiquark, and a gluon.

Quarks and Gluons 2.0: Believing Is Seeing

Of course, a legitimate picture doesn't emerge from the camera with labels "quark, antiquark, gluon" attached. It needs some interpretation.

First, let's take stock of the objects in the picture using everyday language. The complicated-looking bits outline magnets and other components of the accelerator and detector. You can make out a narrow tube running through the middle. That's the beam pipe, through which the electrons and positrons circulate. What's in the picture is only a very small part, a few meters on a side, of the LEP machine, built inside a circular tunnel 27 kilometers in circumference. (By the way, the same tunnel houses the Large Hadron Collider (LHC), which uses protons instead of electrons and positrons, and which operates at higher energies. We'll have much more to say about the LHC in later chapters.) Beams of electrons and positrons, circulating in opposite directions, were accelerated up to enormous energies, until their speed reached within a part in ten billion of the speed of light. The two beams crossed at a few points, where collisions occurred. Those special points were surrounded by big detectors, which could track sparks and capture heat from the particles that emerged from the collisions. The emerging explosion of lines you see are the tracks, and the dots on the outside represent the heat.

The next step is to translate our description of what we see from the language of surface appearance into the language of deep structure. This translation entails such a big conceptual step that you might say it involves a leap of faith.[7] Before taking the leap, let's fortify our faith.

Father James Malley, S.J., taught me a most profound and valuable principle of scientific technique. (It has many other applications as well.) He claimed that he learned this principle at seminary, where it was taught as the Jesuit Credo. It states

> It is more blessed to ask forgiveness than permission.

I'd been following this credo intuitively for years without realizing its ecclesiastical sanction. Now I use it more systematically, and with a clearer conscience.

In theoretical physics, there is a wonderful synergy between the Jesuit Credo and Einstein's "make things simple, but not too simple." Together, they tell us we should make the most optimistic assumptions we dare about how simple things are.[8] If it turns out badly, we can always count on forgiveness and try again—without pausing for permission.

In that spirit, let's make the simplest guess for how to account for what emerges from the collisions, starting from our ideas about the deep structure of the physical world. According to QED an electron and its antiparticle, a positron, can annihilate one another, producing a virtual photon. The virtual photon, in turn, can turn into a quark and an antiquark. So says QED. This core process is shown in Figure 6.3.

---

7. It's not a quantum leap: quantum leaps are small.

8. Of course, "simple" is a complicated concept. See Chapter 12.

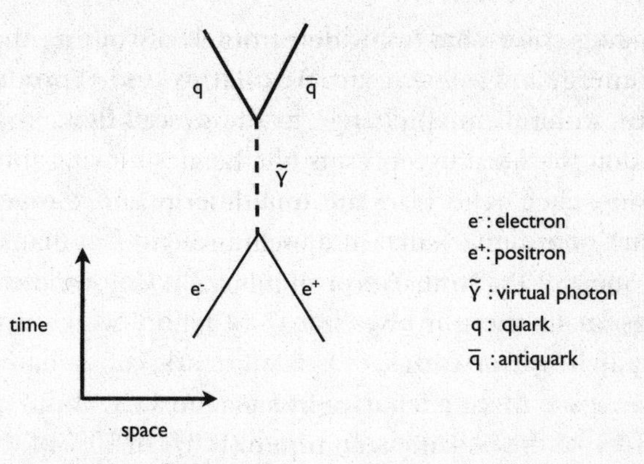

Figure 6.3 A space-time diagram of the core process, in which an electron and positron annihilate into a virtual photon, which then materializes as a quark-antiquark pair.

At that point things get dicey because, as we've discussed, the quark (and antiquark) cannot exist in isolation. They must be confined inside hadrons. The process of acquiring virtual particle clouds and canceling color charges, which leads from quarks to hadrons, might be very complicated. These complications could make it difficult to identify signs of the original quark and antiquark, just as it would be very difficult, looking at the final mess, to figure out which rock started a rockslide. But let's try to think it through, in the spirit of the Credo, hoping for the best.

The initial quark and antiquark that emerge from the collision have enormous energy and are moving in opposite directions.[9] Now suppose that the process of acquiring clouds and canceling color charge is usually accomplished gently, by producing and

9. This is because the total momentum is conserved. It was zero at the start, because the electrons and positrons were moving at the same speed in opposite directions. So it must still be zero at the end, as is observed. Of course, in principle we might discover experimentally that momentum is *not* conserved, but then we'd really need to backtrack, all the way to unlearning first-year physics.

rearranging color charges, without much disrupting the overall flow of energy and momentum. We call this kind of production of particles, without much change in the overall flow, "soft" radiation. Then we'd see two swarms of particles moving in opposite directions, each inheriting the total energy and momentum of the quark or antiquark that initiated it. And in fact that's what we do see, most of the time. A typical picture is Color Plate 2.

Occasionally there is also "hard" radiation, which does affect the overall flow. The quark, or the antiquark, can radiate a gluon. Then we'll see three jets instead of two. At LEP, this happens in about 10% of the collisions. In roughly 10% of 10% of the events (that is, in 1%) there will be four jets, and so on.

The theoretical interpretation of our photographs is sketched in Figure 6.4. With this interpretation, we can eat our quarks and have them too. Even though isolated quarks are never observed, we can see them through the flows they induce. In particular, we can check whether the probabilities for producing different numbers of jets, coming out at different angles and sharing the total energy in different ways, match the probabilities we calculate for quarks, antiquarks, and gluons doing these things in QCD. LEP produced hundreds of millions of collisions, so the comparison between theoretical predictions and experimental results can be done precisely and in great detail.

It works. And that's why I can say with total confidence that the objects you're seeing in Color Plate 1 are a quark, an antiquark, and a gluon. To see these particles, however, we've had to expand our notions of what it means to see something—and of what a particle is.

Let's bring our appreciation of the quark/gluon photographs to a climax by connecting it to two big ideas: asymptotic freedom and quantum mechanics.

There's a direct connection between quarks and gluons appearing as jets and asymptotic freedom. It's easy to explain the connection using Fourier transforms, but unfortunately Fourier transforms themselves aren't so easy to explain, so we won't go

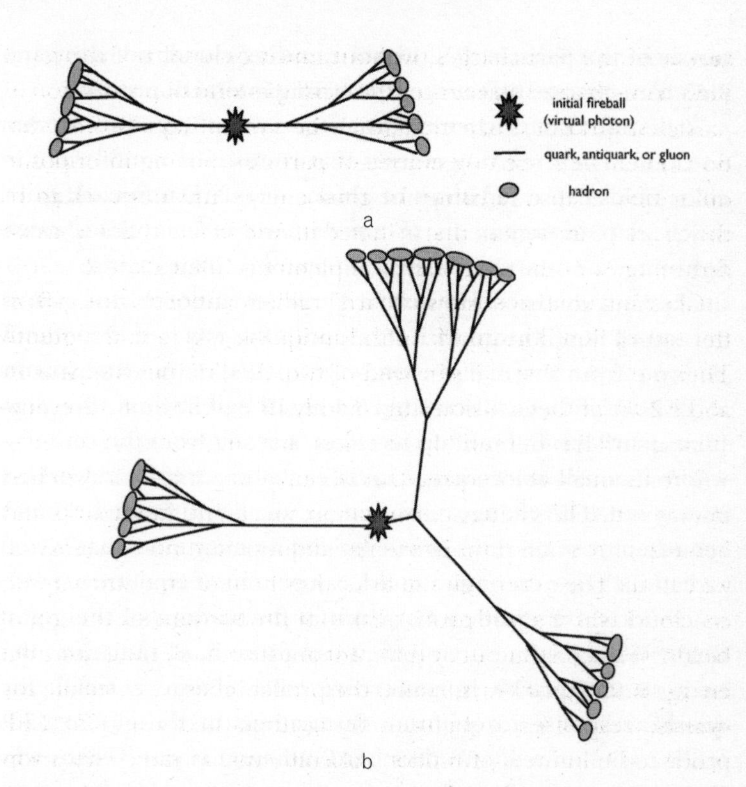

Figure 6.4 a. How soft radiation makes hadron jets out of a quark and antiquark. b. How hard radiation of a gluon, followed by lots of soft radiation, makes three jets.

there. Here's a word explanation that is less precise but calls for more imagination (and less preparation):

To explain why quarks and gluons appear (only) as jets, we have to explain why soft radiation is common but hard radiation is rare. The two central ideas of asymptotic freedom are: first, that the intrinsic color charge of the fundamental particle—whether quark, antiquark, or gluon—is small and not very powerful; second, that the cloud of virtual particles that surrounds the fundamental particle is thin nearby, but grows thicker far away. It's the surrounding cloud that enhances the fundamental power of the particle. It's the surrounding cloud, not the particle's core charge, that makes the strong interaction strong.

Radiation occurs when a particle gets out of equilibrium with its cloud. Then rearrangements that restore the equilibrium in color fields cause radiation of gluons or quark-antiquark pairs, much as rearrangements in atmospheric electric fields cause lightning, or rearrangements in tectonic plates cause earthquakes and volcanos. How can a quark (or antiquark, or gluon) get out of equilibrium with its cloud? One way is if it suddenly pops out from a virtual photon, as happened in the experiments at LEP we've been discussing. To reach equilibrium, the newborn quark has to build up its cloud, starting from the center—where its small color charge initiates the process—and working its way out. The changes involved are small and graded, so they require only small flows of energy and momentum—that is, soft radiation. The other way a quark can get out of equilibrium with its cloud is if it's jostled by quantum fluctuations of the gluon fields. If the jostling is violent, it can cause hard radiation. But because the quark's intrinsic core color charge is small, the quark's response to quantum fluctuations in the gluon fields tends to be limited, and thus hard radiation is rare. That's why three jets are less likely than two.

The connection of our photographs to the profundities of quantum mechanics is even more direct, and needs no such elaborate explanation. It is simply that once more, we find that doing the same thing over and over again gives different results each time. We saw that before with the ultrastroboscopic nanomicroscope that takes pictures of protons; we're seeing it now with the creative destruction machine that takes pictures of empty space. If the world behaved classically and predictably, the billion euros invested in LEP would have underwritten a very boring machine: every collision would just reproduce the result of the first one, and there'd be only one photograph to look at. Instead, our quantum-mechanical theories predict that many results can emerge from the same cause. And that is what we find. We can predict the relative probabilities of different results. Through many repetitions, we can check those predictions in detail. In that way, short-term unpredictability can be tamed. Short-term unpredictability is, in the end, perfectly compatible with long-term precision.

7

Symmetry Incarnate

Color gluons are embodied ideas: symmetry incarnate.

THE CENTRAL IDEA OF QCD IS SYMMETRY. Now *symmetry* is a word that's in common use, and like many such words its meaning is fuzzy at the edges. Symmetry can mean balance, pleasing proportions, regularity. In mathematics and physics, its meaning is consistent with all those ideas, but sharper.

The definition I like is that *symmetry means you have a distinction without a difference.*

The legal profession uses that phrase, distinction without a difference, as well. In that context, it typically means saying the same thing in different ways, or—to put it less politely—quibbling. Here's an example, from the comedian Alan King:

> My lawyer warned me that if I died without a will I'd die intestate.

To understand the mathematical concept of symmetry, it's best to think about an example. We can build a nice little tower of examples, one that contains the most important ideas in easily digestible form, in the world of triangles. (See Figure 7.1.) You can't move most triangles around without changing them (Figure 7.1a). Equilateral triangles, however, are special. You can rotate an equilateral through 120° or 240° (twice as much) and still get the same shape (Figure 7.1b). The equilateral triangle has nontrivial symmetry,

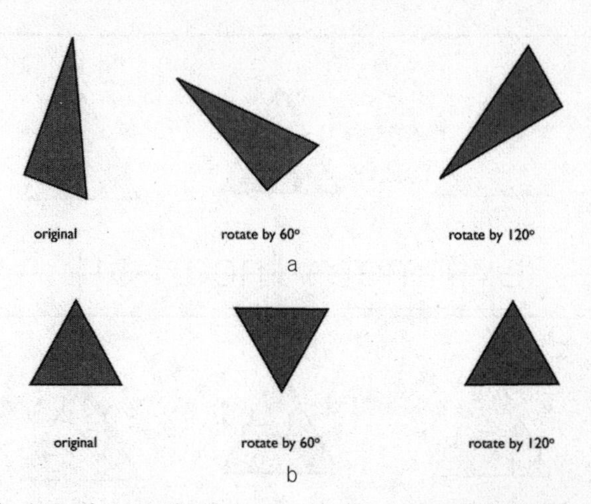

Figure 7.1 A simple example of symmetry. a. You can't move a lopsided triangle without changing it. b. If you rotate an equilateral triangle through 120° around its center, it does not change.

because it permits *distinctions* (between a triangle and its rotated versions) that don't, after all, make any *difference* (the rotated versions give the same shape). Conversely, if someone tells you that a triangle looks the same after rotation by 120°, you can deduce that it's an equilateral triangle (or that the person's a liar).

The next level of complexity comes when we consider a set of triangles with different kinds of sides. (See Figure 7.2.) Now, of course, if we rotate one of the triangles through 120°, we don't get the same thing—the sides don't match. In Figure 7.2, the first triangle (RBG) rotates into the second triangle (BGR), the second triangle rotates into the third (GRB), and the third rotates into the first. But the complete *set*, containing *all three triangles*, is not changed.[1]

Conversely, if someone tells you that a triangle with three different kinds of sides, together with some other stuff, looks the same

1. For this example, you're supposed to ignore the fact that the three triangles are in different places. If that bugs you, you can imagine that they're infinitely thin triangles stacked on top of one another.

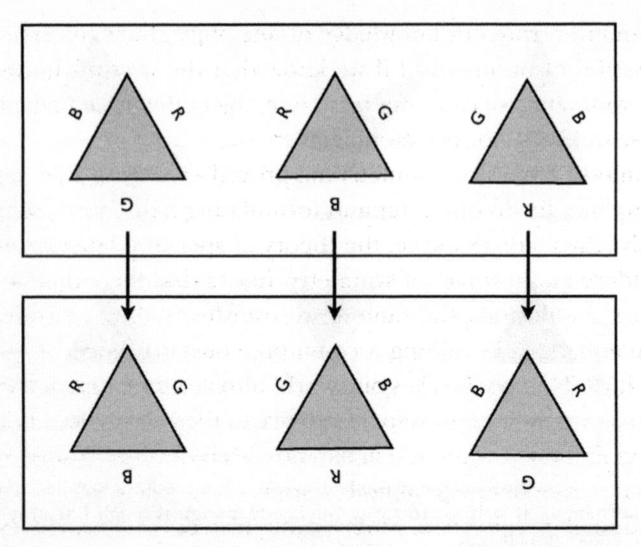

Figure 7.2 A more complicated example of symmetry. Equilateral triangles with different "colored" sides (here the colors are suggested as R[ed], B[lue], G[reen]) change under rotations through 120°; but the entire *set of three* goes over into itself.

after rotation through 120°, you can deduce both that the triangle is equilateral *and* that there are two more equilateral triangles with different arrangements of the sides (or that the person's a liar).

Let's add one last layer of of complexity. Instead of triangles with differently colored sides, let's consider laws involving those triangles. For example, a simple law could be that if you squeeze the triangle it collapses in a neat way, so that its sides become bowed. Now suppose that we had investigated only RBG triangles, so that we really established the squeezing law only for those triangles. If we knew that rotating by 120° was a distinction without a difference—that is, that rotating by 120° defined a symmetry, in the mathematical sense—we'd be able to deduce not only that there had to be the other kinds of triangles, but also that they too collapse neatly when you squeeze them.

This series of examples exhibits, in simple forms, the powers of symmetry. If we know an object has symmetry, we can deduce some of its properties. If we know a set of objects has symmetry, we can infer from our knowledge of one object the existence and properties of others. And if we know that the laws of the world have symmetry, we can infer from one object the existence, properties, and behavior of new objects.

In modern physics, symmetry has proved a fruitful guide to predicting new forms of matter and formulating new, more comprehensive laws. For example, the theory of special relativity can be considered a postulate of symmetry. It says that the equations of physics should look the same if we transform all the entities in those equations by adding a common, constant "boost" to their velocities. That boost takes one world into another, distinct world moving with a constant velocity relative to the first. Special relativity says that that distinction makes no difference—the same equations describe behavior in both worlds.

Although the details are more complicated, the procedures for using symmetry to understand our world are basically the same as the ones we used in our simple examples from triangle world. We consider that our equations can be transformed in ways that could, in principle, change them—and then we demand that they don't in fact change. The possible distinction makes no difference. Just as in the triangle world examples, so in general, for a symmetry to hold several things have to be true. The entities that appear in the equations will need to have special properties, to come in related sets, and to obey tightly related laws.

Thus symmetry can be a powerful idea, rich in consequences. It's also an idea that Nature is very fond of. Prepare for a public display of affection.

Nuts and Bolts, Hubs and Sticks

The theory of quarks and gluons is called quantum chromodynamics, or QCD. The equations of QCD are displayed in Figure 7.3.[2]

---

2. There won't be a quiz.

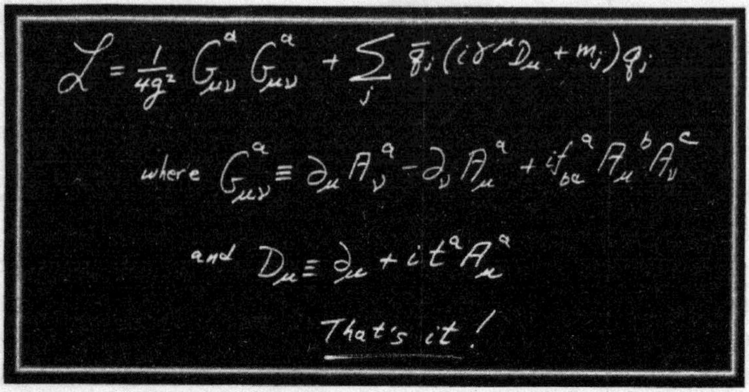

$$\mathcal{L} = \frac{1}{4g^2} G^a_{\mu\nu} G^a_{\mu\nu} + \sum_j \bar{q}_j (i\gamma^\mu D_\mu + m_j) q_j$$

where $G^a_{\mu\nu} \equiv \partial_\mu A^a_\nu - \partial_\nu A^a_\mu + i f^a_{bc} A^b_\mu A^c_\nu$

and $D_\mu \equiv \partial_\mu + i t^a A^a_\mu$

That's it!

Figure 7.3 The QCD Lagrangian $\mathcal{L}$ written out here provides, in principle, a complete description of the strong interaction. Here m_j and q_j are the mass and quantum field of the quark of jth flavor, and A is the gluon field, with space-time indices μ, ν and color indices a, b, c. The values of the numerical coefficients f and t are completely determined by color symmetry. Aside from the quark masses, the coupling constant g is the single free parameter of the theory. In practice, it takes ingenuity and hard work to calculate anything using $\mathcal{L}$.

Pretty compact, no? Nuclear physics, new particles, weird behaviors, the origin of mass—it's all there!

Actually, you shouldn't be too quick to be impressed by the fact that we can write equations in compact form. Our clever friend Feynman demonstrated how to write down the Equation of the Universe in a single line. Here it is:

$$U = 0 \tag{1}$$

U is a definite mathematical function, the total unworldliness. It's the sum of contributions from all the piddling partial laws of physics. To be precise, $U = U_{\text{Newton}} + U_{\text{Einstein}} + \cdots$. Here, for instance, the Newtonian mechanical Unworldiness U_{Newton} is defined by $U_{\text{Newton}} = (F - ma)^2$; the Einstein mass-energy Unworldliness is defined by $U_{\text{Einstein}} = (E - mc^2)^2$; and so forth. Because every contribution is positive or zero, the only way the total U can vanish is for every contribution to vanish, so $U = 0$ implies

$F = ma$, $E = mc^2$, and any other past or future law you care to include!

Thus we can capture all the laws of physics we know, and accommodate all the laws yet to be discovered, in one unified equation. The Theory of Everything!!! But it's a complete cheat, of course, because there is no way to use (or even define) U, other than to deconstruct it into its separate pieces and then use those.

The equations displayed in Figure 7.3 are quite unlike Feynman's satirical unification. Like $U = 0$, the master equations of QCD encode a lot of separate smaller equations. (For experts: the master equations involve matrices of tensors and spinors; the smaller equations, for their components, involve ordinary numbers.) There's a big difference, though. When we unpack $U = 0$, we get a bunch of unrelated stuff. When we unpack the master equations of QCD, we get equations that are related by symmetry—symmetry among the colors, symmetry among different directions in space, and the symmetry of special relativity between systems moving at constant velocity. Their complete content is out front, and the algorithms that unpack them flow from the unambiguous mathematics of symmetry. So let me assure you that you really should be impressed. It's a genuinely elegant theory.

One reflection of that elegance is that the essence of QCD can be portrayed, without severe distortion, in a few simple pictures. They're displayed in Figure 7.5. We'll discuss them presently.

But first, for comparison and as a warm-up, I'd like to display the essence of quantum electrodynamics (QED) in a similar format. QED is, as its name suggests, the quantum-mechanical account of electrodynamics. QED is a slightly older theory than QCD. The basic equations of QED were in place by 1931, but for quite a while people made mistakes in trying to solve them, and got nonsensical (infinite) answers, so the equations got a bad reputation. Around 1950 several brilliant theorists (Hans Bethe, Sin-Itiro Tomonaga, Julian Schwinger, Richard Feynman, and Freeman Dyson) straightened things out.

The essence of QED can be portrayed in the single picture of Figure 7.4a. It shows that a photon responds to the presence or motion of electric charge. Though it looks cartoonish, that little picture is much more than a metaphor. It is the core process in a rigorous representation of a systematic method of solving the equations of QED, which we owe to Feynman. (Yes, him again. Sorry Murray.) Feynman diagrams depict processes in space and time whereby particles travel from given places at one time to other places at some later time. In between, they can influence each other. The possible processes and influences in quantum electrodynamics are built up by connecting world-lines (that is, paths through space and time) of electrons and photons in arbitrary ways using the core process. It's easier done than said, and you'll get the general idea after pondering Figures 7.4b–f.

Perfectly definite mathematical rules specify, for each Feynman diagram, how likely it is for the process it depicts to occur. The rules for complicated processes, perhaps involving many real and virtual charged particles and many real and virtual photons, are built up from the core process. It is like making constructions with TinkerToys. The particles are different kinds of sticks you can use, and the core process provides the hubs that join them. Given these elements, the rules for construction are completely determined. For example, Figure 7.4b shows a way in which the presence of one electron affects another. The rules of Feynman diagrams tell you how likely it is that exchange of one virtual photon, as depicted there, makes the electrons swerve by any specified amount. In other words, they tell you the force! This diagram encodes the classic theory of electric and magnetic forces that we teach to undergraduates. There are corrections to that theory when you take into account rarer processes, involving the exchange of two virtual photons, as shown in Figure 7.4c. Or a photon can break free, as in Figure 7.4d; that's what we call electromagnetic radiation, one form of which is light. You can also have processes where all the particles are virtual, as in Figure 7.4e. None of the particles involved can be observed, so that "vacuum"

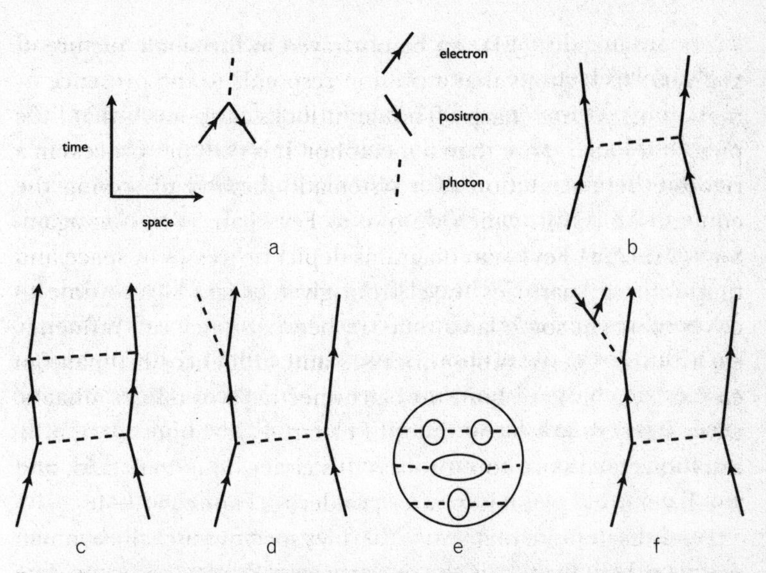

Figure 7.4 a. The essence of QED: photons respond to electric charge. b. A good approximation to the force between electrons, due to exchange of virtual photons. c. A better approximation includes contributions like this. d. Let there be light! An accelerated electron can emit a photon. e. A totally virtual process. f. Radiation of an electron-positron pair. The antielectron, or positron, is represented as an electron with the arrows reversed.

process might seem academic or metaphysical, but we'll see that processes of this kind are tremendously important.[3]

Maxwell's equations for radio waves and light, Schrödinger's equation for atoms and chemistry, and Dirac's more refined version, which includes spin and antimatter—all that and more is faithfully encoded in these squiggles.

In the same pictorial language, QCD appears as an expanded version of QED. Its more elaborate set of ingredients and core

3. I once had a very interesting conversation about that with Feynman himself. He told me that he had originally hoped to remove vacuum processes from the theory, and was very disappointed to find that he couldn't do it in a consistent way. I'll tell you more about that conversation in Chapter 8.

processes are displayed in Figure 7.5, which has a correspondingly more elaborate caption.

At this pictorial level QCD is a lot like QED, but bigger. The diagrams look similar, and the rules for evaluating them are similar, but there are more kinds of sticks and hubs. More precisely, while there is just one kind of charge in QED—namely, electric charge—QCD has three.

The three kinds of charge that appear in QCD are called, for no good reason, *colors.* These "colors" of course have nothing to do with color in the ordinary sense; rather, they are deeply similar to electric charge. In any case, we'll label them red, white, and blue. Every quark has one unit of one of the color charges. In addition, quarks come in different species, or *flavors.* The only two flavors that play a role in normal matter are called *u* and *d,* for up and down.[4] Quark "flavors" have no more to do with how anything tastes than quark colors have to do with how they look. Also, these mixed-metaphorical names for *u* and *d* (Zen koan: What is the taste of up?) do not imply that there is any real connection between flavors and directions in space. Don't blame me; when I get the chance, I give particles dignified scientific-sounding names like axion and anyon.

Continuing the analogy between QED and QCD, there are photon-like particles, called color gluons, that respond in appropriate ways to the presence or motion of color charge, much as photons respond to electric charge.

So there are *u* quarks with a unit of red charge, *d* quarks with a unit of blue charge—six different possibilities altogether. And instead of one photon that responds to electric charge, QCD has eight color gluons that can either respond to different color charges or change one color charge into another. So there are

4. Earlier I mentioned a third quark flavor, the strange quark *s.* There are also three more quark flavors: charm *c,* bottom *b,* and top *t.* These are even heavier and more unstable than *s.* We'll ignore them all for now.

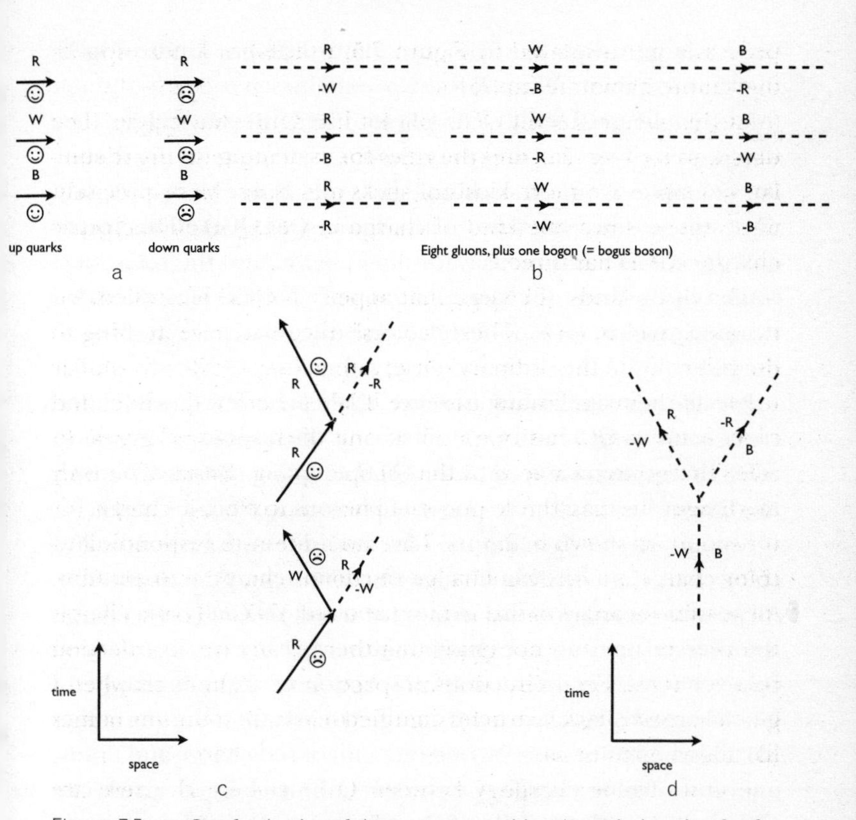

Figure 7.5 a. Quarks (antiquarks) carry one positive (negative) unit of color charge. They play a role in QCD similar to that of electrons in QED. A complication is that there are several distinct kinds, or flavors, of quarks. The two that are important for normal matter are the lightest ones, called *u* and *d*. (To be honest, there are also different flavors of electrons, called muons and τ leptons, but I've been suppressing irrelevant complications.) b. There are 8 different color gluons. Each carries away a unit of color charge and brings in another color (possibly the same). The total of each color charge is conserved. There would seem to be $3 \times 3 = 9$ possibilities for gluons. But one particular combination, the so-called color singlet, which responds equally to all charges, is different from the others. We must remove it if we are to have a perfectly symmetric theory. Thus we predict that exactly 8 gluons exist. Fortunately, that conclusion is vindicated by experiment. Gluons play a role in QCD similar to that of photons in QED. c. Two representative core processes, where gluons simply respond to, or both respond to and transform, the color charge of quarks. d. A qualitatively new feature of QCD, compared to QED, is that there are processes whereby color gluons respond to one another. Photons do not.

quite a large variety of sticks, and many different kinds of hubs that connect them. With all these possibilities, it seems as though things could get terribly complicated and messy. And so they would, were it not for the overwhelming symmetry of the theory. For example, if you interchange red with blue everywhere, you must still get the same rules. The symmetry of QCD allows you to mix the colors continuously, forming blends, and the rules must come out the same for blends as for pure colors. This extended symmetry is extremely powerful. It fixes the relative strength of all the hubs.

For all their similarities, however, there are a few crucial differences between QCD and QED. First of all, the response of gluons to color charge, as measured by the QCD coupling constant, is much more vigorous than the response of photons to electric charge.

Second, as shown in Figure 7.5c, in addition to responding to color charge, gluons can change one color charge into another. All possible changes of this kind are allowed. Yet each color charge is conserved, because the gluons themselves can carry unbalanced color charges. For example, if absorption of a gluon changes a blue-charged quark into a red-charged quark, then the gluon that got absorbed must have carried one unit of red charge and minus one unit of blue charge. Conversely, a blue-charged quark can emit a gluon with one unit of blue charge and minus one unit of red charge; it becomes a red-charged quark.

A third difference between QCD and QED, the most profound, is a consequence of the second. Because gluons respond to the presence and motion of color charge, and gluons carry unbalanced color charge, it follows that gluons, quite unlike photons, respond directly to one another.

Photons, by contrast, are electrically neutral. They don't respond to one another very much at all. We're all familiar with this, even if we've never thought much about it. When we look around on a sunny day, there is reflected light bouncing every which way, but we see right through it. The laser sword fights you've seen in *Star Wars* films wouldn't work. (Possible explanation: it's a movie about a technologically advanced civilization in a galaxy far, far away, so maybe they're using color gluon lasers.)

Each of these differences makes calculating the consequences of QCD more difficult than calculating the consequences of QED. Because the basic coupling is stronger in QCD than in QED, more complicated Feynman diagrams, with lots of hubs on the inside, make relatively larger contributions to any process. And because there are various possibilities for routing the flows of color, and more kinds of hubs, there are many more diagrams at each level of complexity.

Asymptotic freedom allows us to calculate some things, such as the overall flows of energy and momentum in jets. That's because the many "soft" radiation events don't much affect the overall flow, and we can ignore them in our calculations. Only the small number of hubs where "hard" radiation occurs demand attention. Without too much work, then, using pencil and paper, a human being can work out predictions for the relative probability of different numbers of jets coming out at different angles with different shares of energy. (It helps if you give the human being a laptop and send her through a few years of graduate school.) In other cases the equations have been solved—approximately—only after heroic efforts. We'll discuss the heroic efforts that enable us to calculate the mass of the proton, starting from massless quarks and gluons—and thus to identify the origin of mass—in Chapter 9.

Quarks and Gluons 3.0: Symmetry Incarnate

In trying to do justice to the way in which postulating an enormous amount of symmetry—what we call *local* symmetry—forces us to include color gluons in our equations, and thus to predict their existence and all their properties, I'm reminded of one of my favorite Piet Hein *Grooks*:

lovers meander in prose and rhyme,
trying to say—

for the thousandth time—
what's easier done than said.

Anyway, on to the prose and rhyme.

Back when we discussed the colored triangles and their symmetry, there was a fussy footnote about how you were supposed to ignore the fact that the different triangles were in different places. It makes perfectly good logical and mathematical sense to do that. In mathematics we often ignore irrelevant details in order to concentrate on the most interesting, essential features. For example, in geometry it's standard operating procedure to deal, conceptually, with lines that have zero thickness and continue forever in both directions. But from the point of view of physics, it's a little peculiar to suppose that symmetry requires you to take no account of where things are. Specifically, for example, it seems peculiar to suppose that the symmetry between red color charge and blue color charge requires that you change red charged quarks into blue charged quarks, and *vice versa*, everywhere in the universe. It might seem more natural to suppose that you could make the changes only locally, without having to worry about distant parts of the universe.

That physically natural version of symmetry is called *local* symmetry. Local symmetry is a much bigger assumption than the alternative, global symmetry. For local symmetry is an enormous collection of separate symmetries—roughly speaking, a separate symmetry for each point of space and time. In our example, we can make the red-blue charge switch anywhere and at any moment. Each place and moment defines, therefore, its own symmetry. With a global symmetry you have to make the same switch everywhere and everywhen, and instead of infinitely many independent symmetries you have just one lockstep version.

Because local symmetry is a much bigger assumption than global symmetry, it imposes more restrictions on the equations or, in other words, on the form of physical laws. The restrictions from local symmetry are so severe, in fact, that at first sight they appear impossible to reconcile with the ideas of quantum mechanics.

Before explaining the problem, a quick summary of the relevant quantum mechanics: In quantum mechanics, we have to allow for the possibility that a particle might be observed at different places, with different probabilities. There's a wave function that describes all these possibilities. The wave function has large values where the probability is large, and small values where the probability is small (quantitatively, the probability is equal to the square of the wave function). Furthermore, wave functions that are nice and smooth—that vary gently in space and time—have lower energy than those that have abrupt changes.

Now to the heart of the problem: Let's suppose we have a nice smooth wave function for a quark carrying red color charge. Then apply our example of local symmetry in a small region, changing red color charge into blue color charge. After that change, our wave function has abrupt changes. Inside the small region, it has only a blue color component; outside, it has only a red color component. So we've changed a low-energy wave function, without abrupt changes, into a wave function that has abrupt changes and therefore describes a state of high energy. That change of state is going to change the behavior of the quark we're describing, unmistakably, because there are many detectable effects of energy. For example, according to Einstein's second law, you could determine the energy of the quark by weighing it. But the whole point of symmetry is that it's not supposed to change the behavior of the things it transforms.[5] We want to have a distinction *without* a difference.

---

5. The boost symmetry of special relativity changes the energy of particles—but it also changes the behavior of the scales you'd use to weigh them, in just such a way that there's no net detectable effect. Our local color symmetry, on the contrary, makes no change in normal scales (such as you find in grocery stores), which have zero overall color charge. So they will register the changed weight, and we're stuck with it.

So to get equations that have local symmetry, we must alter the rule according to which abrupt changes in the wave function necessarily have large energy. We have to suppose that the energy is not governed simply by the steepness of change in the wave function; it must contain additional correction terms. That's where the gluon fields come in. The correction term contains products of various gluon fields (eight of them, for QCD) with the different color components of the quark wave functions. If you do things just right, then when you make a local symmetry transformation, the quark wave function changes, and the gluon field changes, but the energy of the wave function—including the correction terms—stays the same. There's no ambiguity about the procedure: local symmetry dictates what you must do, every step of the way.

The details of the construction are very hard to convey in words. It really is, as the Grook says, "easier done than said," and if you want to see it done for real, with equations, you'll have to look at technical articles or textbooks. I've mentioned a few of the more accessible ones in the endnotes. Fortunately, you don't need to work through the detailed construction to understand the grand philosophical point, which is this:

In order to have local symmetry, we must introduce the gluon fields. And we must arrange the way those gluons fields interact with quarks, and with one another, just so. An *idea*—local symmetry—is so powerful and restrictive that it produces a definite set of equations. In other words, implementing an idea leads to a candidate reality.

The candidate reality containing color gluons succeeds in embodying the idea of local symmetry. New ingredients—color gluon fields—are part of the recipe for its candidate world. Are they present in our world? As we've discussed, and even seen in photographs, they are indeed. The candidate reality, hatched from ideas, is reality itself.

8

The Grid (Persistence of Ether)

What is space? An empty stage, where the physical world of matter acts out its drama? An equal participant, that both provides background and has a life of its own? Or the primary reality, of which matter is a secondary manifestation? Views on this question have evolved, and several times changed radically, over the history of science. Today, the third view is triumphant. Where our eyes see nothing, our brains, pondering the revelations of sharply tuned experiments, discover the Grid that powers physical reality.

Philosophical and scientific ideas about what the world is made of continue to change. Many loose ends remain in today's best world-models, and some big mysteries. Clearly the last word has not been spoken. But we know a lot, too—enough to draw some surprising conclusions that go beyond piecemeal facts. They address, and offer some answers to, questions that have traditionally been regarded as belonging to philosophy or even theology.

For natural philosophy, the most important lesson we learn from QCD is that what we perceive as empty space is in reality a powerful medium whose activity molds the world. Other developments in modern physics reinforce and enrich that lesson. Later, as we explore the current frontiers, we'll see how the concept of "empty" space as a rich, dynamic medium empowers our best thinking about how to achieve the unification of forces.

So: What is the world made of? Subject, as ever, to addition and correction, here is the multifaceted answer that modern physics provides:

- The primary ingredient of physical reality, from which all else is formed, fills space and time.
- Every fragment, each space-time element, has the same basic properties as every other fragment.
- The primary ingredient of reality is alive with quantum activity. Quantum activity has special characteristics. It is spontaneous and unpredictable. And to observe quantum activity, you must disturb it.
- The primary ingredient of reality also contains enduring material components. These make the cosmos a multilayered, multicolored superconductor.
- The primary ingredient of reality contains a metric field that gives space-time rigidity and causes gravity.
- The primary ingredient of reality weighs, with a universal density.

There are words that capture different aspects of this answer. *Ether* is the old concept that comes closest, but it bears the stigma of dead ideas and lacks several of the new ones. *Space-time* is logically appropriate to describe something that is unavoidably *there,* everywhere and always, with uniform properties throughout. But *space-time* carries even more baggage, including a heavy suggestion of emptiness. *Quantum field* is a technical term that summarizes the first three aspects, but it doesn't include the last three and it sounds, well, too technical and forbidding for use in natural philosophy.

I will use the word *Grid* for the primary world-stuff. That word has several advantages:

- We're accustomed to using mathematical grids to position layers of structure, as in Figure 8.1.

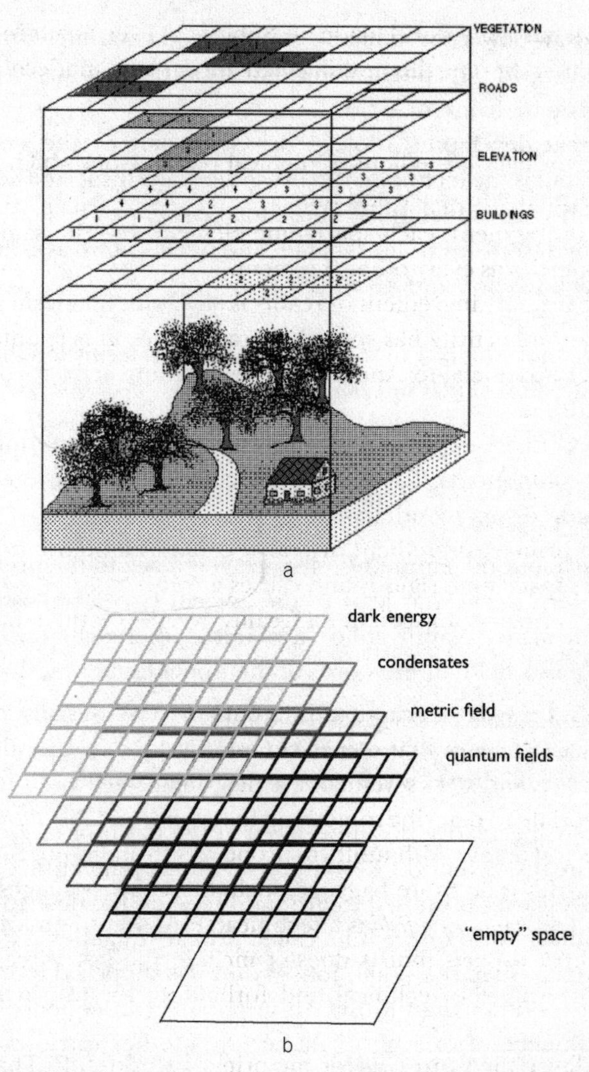

Figure 8.1 Grid, old and new. a. A grid is often used to describe how various things are distributed in space. b. The Grid, which underlies our most successful world-model, has several aspects. The Grid, with these aspects, is present always and everywhere. Ordinary matter is a secondary manifestation of the Grid, tracing its level of excitation.

- We draw power for appliances, lights, and computers from the electric grid. The physical world of appearance draws its power, in general, from the Grid.
- A great developing project, driven in part by the needs of physics,[1] is the technology to integrate many dispersed computers into functional units, whose total power can be accessed as needed from any point. That technology is known as Grid technology. It's hot, and it's cool.
- *Grid* is short.
- *Grid* is not *Matrix*. I'm sorry, but the sequels tarnished that candidate. And *Grid* is not *Borg*.

A Brief History of Ether

Debate about the emptiness of space goes back to the prehistory of modern science, at least to the ancient Greek philosophers. Aristotle wrote, "Nature abhors a vacuum," whereas his opponents the atomists held, in the words of their poet Lucretius, that

> All nature, then, as self-sustained, consists
> Of twain of things: of bodies and of void
> In which they're set, and where they're moved around.

That old speculative debate echoed at the dawn of modern science, in the Scientific Revolution of the seventeenth century. René Descartes proposed to ground the scientific description of the natural world on what he called primary qualities: extension (essentially, shape) and motion. Matter was supposed to have no properties other than those. An important consequence is that the influence of one bit of matter on another can occur only through contact; having no properties other than extension and motion, a bit of matter has no way of knowing about other bits other than by touching them. Thus, to describe (for instance) the

1. As we'll discuss later.

motion of planets, Descartes had to introduce an invisible space-filling "plenum" of invisible matter. He envisioned a complex sea of whirlpools and eddies, upon which the planets surf.

Isaac Newton cut through all those potential complexities by formulating precise, successful mathematical equations for the motion of planets, using his laws of motion and of gravity. Newton's law of gravity does not fit into Descartes's framework. It postulates action at a distance, rather than influence by contact. For example, the Sun exerts a gravitational force on Earth, according to Newton's law, even though it is not in contact with Earth. Despite the success of his equations in providing an excellent, detailed account of planetary motion, Newton was not happy with action at a distance:

> That one body may act upon another at a distance through a vacuum without the mediation of anything else, by and through which their action and force may be conveyed from one another, is to me so great an absurdity that, I believe, no man who has in philosophic matters a competent faculty of thinking could ever fall into it.

Nevertheless he left his equations to speak for themselves:

> I have not been able to discover the cause of those properties of gravity from phenomena, and I frame no hypotheses; for whatever is not deduced from the phenomena is to be called a hypothesis, and hypotheses, whether metaphysical or physical, whether of occult qualities or mechanical, have no place in experimental philosophy.

Newton's followers, of course, did not fail to notice that his system had emptied out space. Having fewer scruples, they became more Newtonian than Newton. Here is Voltaire:

> A Frenchman who arrives in London, will find philosophy, like everything else, very much changed there. He had left the world a plenum, and he now finds it a vacuum.

Mathematicians and physicists, through familiarity and spectacular success, grew comfortable with action at a distance. So things stood, in essence, for more than 150 years. Then James Clerk

Maxwell, consolidating everything that was known about electricity and magnetism, found that the resulting equations were inconsistent. In 1864, Maxwell found that he could repair the inconsistency by introducing an extra term into the equations—in other words, by postulating the existence of a new physical effect. Some years earlier, Michael Faraday had discovered that when magnetic fields change in time, they produce electric fields. To fix his equations, Maxwell had to postulate the converse effect: that changing electric fields produce magnetic fields. With this addition, the fields can take on a life of their own. Changing electric fields produce (changing) magnetic fields, which produce changing electric fields, and so forth, in a self-renewing cycle.

Maxwell found that his new equations—known today as Maxwell's equations—have pure-field solutions of this kind, solutions that move through space at the speed of light. Climaxing a grand synthesis, he concluded that these self-renewing disturbances in electric and magnetic fields *are* light, a conclusion that has stood the test of time. For Maxwell, these fields that fill all space and take on a life of their own were a tangible symbol of God's glory:

> The vast interplanetary and interstellar regions will no longer be regarded as waste places in the universe, which the Creator has not seen fit to fill with the symbols of the manifold order of His kingdom. We shall find them to be already full of this wonderful medium; so full, that no human power can remove it from the smallest portion of Space, or produce the slightest flaw in its infinite continuity.

Einstein's relationship with the ether was complex, and changed over time. It is also, I think, poorly understood, even by his biographers and historians of science (or quite possibly by me). In his first 1905 paper on special relativity,[2] "On the Electrodynamics of Moving Bodies," he wrote,

---

2. In his second paper, he derived Einstein's second law.

> The introduction of a "luminiferous ether" will prove to be superfluous inasmuch as the view here to be developed will not require an "absolutely stationary space" provided with special properties, nor assign a velocity-vector to a point of the empty space in which electromagnetic processes take place.

That forceful declaration from Einstein puzzled me for a long time, for the following reason. In 1905, the problem facing physics was not that there was no theory of relativity. The problem was that there were *two mutually inconsistent* theories of relativity. On one hand was the relativity theory of mechanics, obeyed by Newton's equations. On the other was the relativity theory of electromagnetism, obeyed by Maxwell's equations.

Both of these relativity theories showed that their respective equations display boost symmetry—that is, the equations take the same form when you add a common, overall velocity to everything. In more physical terms, the laws of physics (as stated by the equations) look the same to any two observers moving at a constant velocity relative to one another. To get from one observer's account of the world to the other, however, you have to relabel positions and times. An observer on a plane from New York to Chicago, for instance, would in a couple of hours label Chicago as "distance 0," whereas Chicago would still be "distance 500 miles west" (roughly) for the observer on the ground. The problem was that the relabeling required for mechanical relativity was different from that required for electromagnetic relativity. According to mechanical relativity you must relabel spatial positions but not times; whereas according to electromagnetic relativity you had to relabel both, in a rather more complicated way that mixes space and time together. (The equations of electromagnetic relativity had, by 1905, already been derived by Hendrik Lorentz and perfected by Henri Poincaré; today they are known as Lorentz transformations.) Einstein's great innovation was to assert the primacy of electromagnetic relativity, and work out the consequences for the rest of physics.

So it was the venerable theory of Newtonian mechanics, not the upstart theory of electromagnetism, that required modification.

The theory based on particles moving through empty space gave way, not the theory based on continuous, space-filling fields. Maxwell's field equations were not modified by the special theory of relativity; on the contrary, they supplied its foundation. One still had the space-filling, potentially self-regenerating electric and magnetic fields that sent Maxwell into raptures. Indeed, the ideas of special relativity almost *require* space-filling fields and, in that sense, explain why they exist, as we'll discuss momentarily.

Why, then, did Einstein express himself so strongly to the contrary? True, he had undermined old ideas about a mechanical ether, made of particles following Newton's laws—in fact he had undermined those laws altogether. But far from eliminating space-filling fields, his new theory elevated their status. He might have said with more justice (I've always thought) that the idea of an ether that looks different to moving observers is wrong but a reformed ether, that looks the same to observers moving at a constant velocity relative to one another, is the natural setting for special relativity.

At the time he was hatching special relativity, in 1905, Einstein was also brooding over the problem of what later became known as light quanta. A few years earlier, in 1899, Max Planck had put forward the first idea of what eventually became quantum mechanics. Planck suggested that atoms could exchange energy with electromagnetic fields—that is, emit and absorb electromagnetic radiation, such as light—only in discrete units, or quanta. Using that idea, he was able to explain some experimental facts about blackbody radiation. (Very roughly speaking, the problem is how the color of a hot body, such as a red-hot poker or a glowing star, depends on its temperature. Less rough, but still far from smoothly polished: a hot body emits a whole range of colors, with different intensities. The challenge was to describe the whole spectrum of intensities and how it changes with temperature.) Planck's idea worked, empirically, but it wasn't very satisfactory intellectually. It was just tacked on to the other laws of physics, not derived from them. In fact, as Einstein (but not Planck) clearly realized, Planck's idea was *inconsistent* with the other laws.

In other words, Planck's idea was another of those things—like the original quark model, or partons—that works in practice but not in theory. It wouldn't pass muster at the University of Chicago, and it didn't pass muster with Einstein. But Einstein was very impressed with the power of Planck's idea to explain experimental results. He extended it in a new direction, hypothesizing that not only did atoms emit and absorb light (and electromagnetic radiation generally) in discrete units of energy, but light always came in discrete units of energy, and also travelled carrying discrete units of momentum. With these extensions, Einstein was able to explain more facts, and even to predict new ones—including the photoelectric effect, which was the primary work cited in his 1921 Nobel Prize. In his mind, Einstein had cut the Gordian knot: Planck's idea is inconsistent with existing physical laws, but it works—therefore those laws must be wrong!

And if light travels in lumps of energy and momentum, what could be more natural than to consider these lumps—and light itself—to be particles of electromagnetism? Fields might be more convenient, as we'll see, but Einstein was never one to value convenience over principle. With this issue occupying his mind, I suspect that Einstein took an unusual perspective on what lessons to draw from special relativity. For him, the idea of a space-filling entity that looks the same when you move past it at finite velocity (as special relativity shows the "luminiferous ether" must) was counterintuitive and therefore suspect. This perspective, which cast a shadow over Maxwell's electromagnetic field theory of light, reinforced his intuitions from Planck's work, and from his own, on blackbody radiation and the photoelectric effect. Einstein thought that these developments—the ether had become counterintuitive, and it seemed to take form, physically, only in lumps—together made a strong case for abandoning fields and going back to particles.

In a 1909 lecture Einstein speculated publicly along these lines:

> Anyway, this conception seems to me the most natural: that the manifestation of light's electromagnetic waves is constrained at singularity points, like the manifestation of electrostatic fields in the

> theory of the electron. It can't be ruled out that, in such a theory, the entire energy of the electromagnetic field could be viewed as localized at these singularities, just like the old theory of action-at-a-distance. I imagine to myself, each such singular point surrounded by a field that has essentially the same character of a plane wave, and whose amplitude decreases with the distance between the singular points. If many such singularities are separated by a distance small with respect to the dimensions of the field of one singular point, their fields will be superimposed, and will form in their totality an oscillating field that is only slightly different from the oscillating field in our present electromagnetic theory of light.

In other words, by 1909—and even, I suspect, in 1905—Einstein did *not* think that Maxwell's equations expressed the deepest reality of light. He did not think the fields truly had a life of their own; instead, they arose from "singularity points." He did not think that they truly filled space: they are concentrated in packets near the singularity points. These ideas of Einstein were, of course, tied up with his concept that light comes in discrete units: today's *photons.*

Just as Newton had misgivings about the natural implication of his theory, that it emptied space, Einstein had misgivings about the natural implication of his theory, that it filled space. Like Columbus, who found the New World while seeking a way to the Old, explorers who make landfall on unexpected continents of ideas are often unprepared to accept what they have found. They continue to seek what they were looking for.

By 1920, after he developed the theory of general relativity, Einstein's attitude had changed: "More careful reflection teaches us, however, that the special theory of relativity does not compel us to deny ether." Indeed, the general theory of relativity is very much an "ethereal" (that is, ether-based) theory of gravitation. (I'm saving Einstein's own declaration on that score for use later in this chapter.) Nevertheless, Einstein never gave up on eliminating the electromagnetic ether:

> If we consider the gravitational field and the electromagnetic field from the standpoint of the ether hypothesis, we find a remarkable difference between the two. There can be no space nor any part of

> space without gravitational potentials; for these confer upon space its metrical qualities, without which it cannot be imagined at all. The existence of the gravitational field is inseparably bound up with the existence of space. *On the other hand a part of space may very well be imagined without an electromagnetic field. . . .*[3]

Around 1982, I had a memorable conversation with Feynman at Santa Barbara. Usually, at least with people he didn't know well, Feynman was "on"—in performance mode. But after a day of bravura performances he was a little tired, and eased up. Alone for a couple of hours, before dinner, we had a wide-ranging discussion about physics. Our conversation inevitably drifted to the most mysterious aspect of our model of the world—both in 1982 and today—the subject of the cosmological constant. (The cosmological constant is, essentially, the density of empty space. Anticipating a little, let me just mention that a big puzzle in modern physics is why empty space weighs so little even though there's so much to it.)

I asked Feynman, "Doesn't it bother you that gravity seems to ignore all we have learned about the complications of the vacuum?" To which he immediately responded, "I once thought I'd solved that one."

Then Feynman became wistful. Ordinarily he would look you right in the eye, and speak slowly but beautifully, in a smooth flow of fully formed sentences or even paragraphs. Now, however, he gazed off into space; he seemed transported for a moment, and said nothing.

Gathering himself again, Feynman explained that he had been disappointed with the outcome of his work on quantum electrodynamics. It was a startling thing for him to say, because that brilliant work was what brought Feynman graphs to the world, as well as many of the methods we still use to do difficult calculations in quantum field theory. It was also the work for which he won the Nobel Prize.

---

3. Italics added.

Feynman told me that when he realized that his theory of photons and electrons is mathematically equivalent to the usual theory, it crushed his deepest hopes. He had hoped that by formulating his theory directly in terms of paths of particles in space-time—Feynman graphs—he would avoid the field concept and construct something essentially new. For a while, he thought he had.

Why did he want to get rid of fields? "I had a slogan," he said. Racheting up the volume and his Brooklyn[4] accent, he intoned it:

> The vacuum doesn't weigh anything [dramatic pause] *because there's nothing there!*

Then he smiled, seemingly content, but subdued. His revolution didn't quite come off as planned, but it was a damned good try.

Special Relativity and the Grid

The theory of special relativity, historically, came out of the study of electricity and magnetism, which culminated in Maxwell's field theory. Thus special relativity arose from a description of the world based on the concept of entities—the electric and magnetic fields—that fill all space. That sort of description was a sharp break from the world-model inspired by Newton's classical mechanics and gravity theory, which dominated earlier thinking. Newton's world-model was based on particles that exert forces on one another through empty space.

The claims of special relativity, however, go beyond electromagnetism. The essence of special relativity is a postulate of symmetry: the laws of physics should take the same form after you boost everything appearing in them by the same, constant velocity. This postulate is a universal claim, grown beyond its electromagnetic roots. That is, the boost symmetry of special relativity applies to *all* the laws of physics. As we noted above, Einstein had to change

4. Actually, deep Queens; Feynman was from Far Rockaway.

Newton's laws of mechanics so that they obeyed the same boost symmetry as electromagnetism.

While the ink was drying on special relativity, Einstein started looking for a way to include gravity in the new framework. It was the beginning of a ten-year search, of which Einstein later said,

> . . . the years of searching in the dark for a truth that one feels, but cannot express; the intense desire and the alternations of confidence and misgiving, until one breaks through to clarity and understanding, are only known to him who has himself experienced them.

In the end he produced a *field*-based theory of gravity, general relativity. We'll have much more to say about that theory later in this chapter. Several other clever people, including notably Poincaré, the great German mathematician Hermann Minkowski, and the Finnish physicist Gunnar Nordström, were also in the hunt, trying to construct theories of gravity consistent with the concepts of special relativity. All were led to field theories.

There's a good general reason to expect that physical theories consistent with special relativity will have to be field theories. Here it comes:

A major result of the special theory of relativity is that there is a limiting velocity: the speed of light, usually denoted *c*. The influence of one particle on another cannot be transmitted faster than that. Newton's law for the gravitational force, according to which the force due to a distant body is proportional to the inverse square of its distance *right now*, does not obey that rule, so it is not consistent with special relativity. Indeed the concept "right now" itself is problematic. Events that appear simultaneous to a stationary observer will not appear simultaneous to an observer moving at constant velocity. Overthrowing the concept of a universal "now" was, according to Einstein himself, by far the most difficult step in arriving at special relativity:

> [A]ll attempts to clarify this paradox satisfactorily were condemned to failure as long as the axiom of the absolute character of times, viz., of simultaneity, unrecognizedly was anchored in the

> unconscious. Clearly to recognize this axiom and its arbitrary character really implies already the solution of the problem.

This is fascinating stuff, but it is well covered in dozens of popular books on relativity, so I won't go further into it here. For present purposes, what's important is simply that there's a limiting velocity, *c*.

Now consider Figure 8.2. In Figure 8.2a, we have the world-lines of several particles. Their positions in space are indicated on the horizontal axis, and the value of time is indicated on the vertical axis. As time progresses, the particles' positions change. The positions for any one particle, followed through time, make that particle's world-line. Of course we should really have three spatial dimensions, but even two are too many to fit on a flat page, and fortunately one is enough to make our point. In Figure 8.2b, you see that if influence propagates at a finite speed, then the influence of particle A (say) on particle B depends on where particle A was in the past. So to get the total force on a particle, we have to sum up the influences from all the other particles, coming from different earlier times. This leads to a complicated description, as emphasized in Figure 8.2b. An alternative, also shown in Figure 8.2c, is to forget about keeping track of the individual past positions, and instead focus on the total influences. In other words, we keep track of a *field* representing the total influence.

That move from a particle description to a field description will be especially fruitful if the fields obey simple equations, so that we can calculate the future values of fields from the values they have now, without having to take past values into account. Maxwell's theory of electromagnetism, general relativity, and QCD all have this property. Evidently, Nature has taken the opportunity to keep things relatively[5] simple by using fields.

---

5. No pun intended. Nope, no pun here, folks.

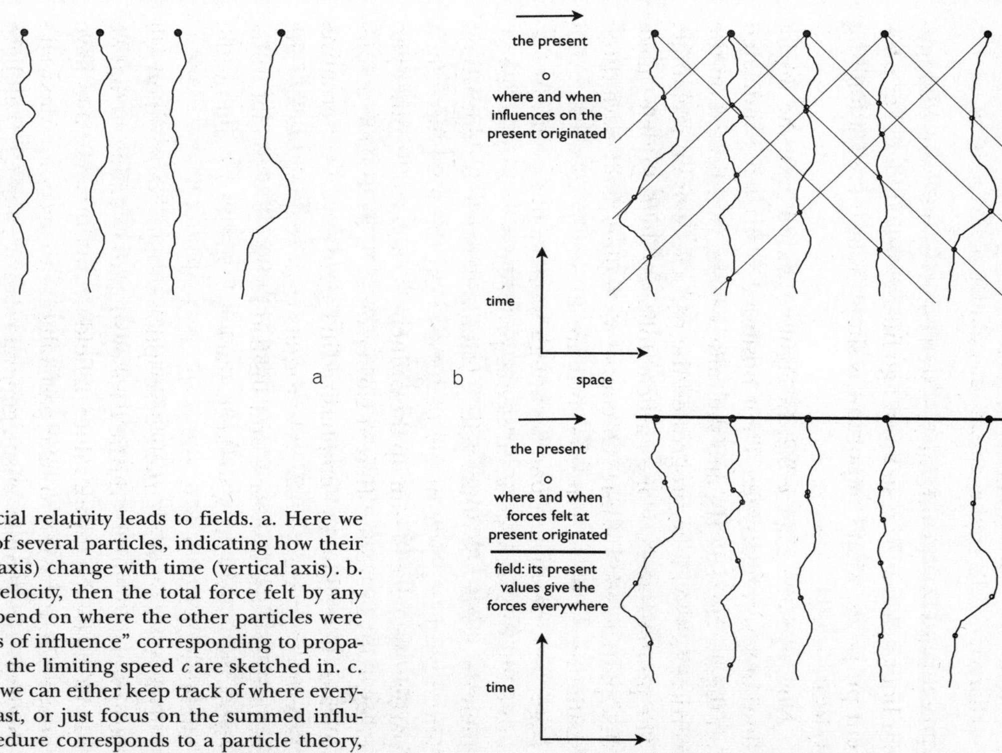

Figure 8.2 How special relativity leads to fields. a. Here we have the world-lines of several particles, indicating how their positions (horizontal axis) change with time (vertical axis). b. If there's a limiting velocity, then the total force felt by any given particle will depend on where the other particles were in the past. The "lines of influence" corresponding to propagation of influence at the limiting speed *c* are sketched in. c. To get the total force, we can either keep track of where everybody's been in the past, or just focus on the summed influences. The first procedure corresponds to a particle theory, the second—potentially much simpler—to a field theory.

Gluons and the Grid

Einstein and Feynman were not unaware of the logic that suggests the inevitability of a field description for fundamental physics. Yet as we've seen, each of them was ready—even eager—to go back to a particle description.

That these two great physicists, at different times and for different reasons, could question the existence of fields that fill all space (a crucial aspect of the Grid) shows that the case for their existence did not appear overwhelming even well into the twentieth century. There was room for doubt, because hard evidence that fields have a life of their own was scanty. In my arguments around Figure 8.2, I made the case that fields are *convenient.* That's very different from their being *necessary ingredients of ultimate reality.*

I'm not sure that Einstein was ever convinced about the electromagnetic ether. One of his greatest strengths as a theoretical physicist could also be a weakness: his stubbornness. Stubbornness served him well when he insisted on resolving the contradictions between the two relativities, mechanical versus electromagnetic, in favor of the latter; again when he insisted on taking Planck's ideas seriously and extending them, despite their conflict with existing theory; and again when he struggled with the difficult and unfamiliar mathematics needed for general relativity. On the other hand, stubbornness kept him from participating in the tremendous successes of modern quantum theory after 1924, when uncertainty and indeterminism took root, and it kept him from accepting one of the most dramatic consequences of his own theory of general relativity, the existence of black holes.

Einstein's difficulties in reconciling the quantum discreteness of photons with continuous space-filling fields, which since Maxwell have been used with great success to describe light, are overcome in the modern concept of quantum fields. Quantum fields fill all space, and the quantum electric and magnetic fields obey Maxwell's equations.[6] Nevertheless, when you observe the quan-

6. That is, to a good first approximation.

tum fields, you find their energy packaged in discrete units: photons. I'll have much more to say about the strange but very successful concepts at the root of quantum field theory in the next chapter.

As for Feynman, he gave up when, as he worked out the mathematics of his version of quantum electrodynamics, he found the fields, introduced for convenience, taking on a life of their own. He told me he lost confidence in his program of emptying space when he found that both his mathematics and experimental facts required the kind of *vacuum polarization* modification of electromagnetic processes depicted—as he found it, using Feynman graphs—in Figure 8.3. Figure 8.3a corresponds to a sophisticated way of summarizing the same physics we saw in Figure 8.2. Here the influence of one particle on another is conveyed by the photon. Figure 8.3b adds something new. Here the electromagnetic field gets modified by its interaction with a spontaneous fluctuation in the electron field—or, in other words, by its interaction with a virtual electron-positron pair. In describing this process, it becomes very difficult to avoid reference to space-filling fields.

The virtual pair is a consequence of spontaneous activity in the electron field. It can occur anywhere. And wherever it occurs, the electromagnetic field can sense it. Those two activities—fluctuations both occurring everywhere and being sensed everywhere—appear quite directly in the mathematical expressions that go with Figure 8.3b. They lead to complicated, small but very specific modifications of the force you would calculate from Maxwell's equations. Those modifications have been observed, precisely, in accurate experiments.

In QED vacuum polarization is a small effect, both qualitatively and quantitatively. In QCD, by contrast, it is all-important. In Chapter 6, we saw how it leads to asymptotic freedom and thereby permits a successful description of jet phenomena. In the next chapter, we'll see how QCD is used to calculate the mass of protons and other hadrons. Our eyes were not evolved to resolve the tiny times (10^{-24} second) and distances (10^{-14} centimeter) where

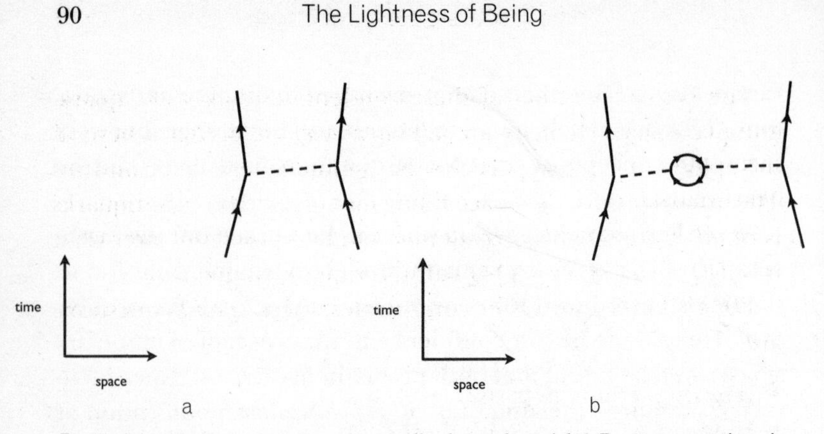

Figure 8.3 The force between electrically charged particles. Part a summarizes, in the language of Feynman graphs, the physics of Figure 8.2. At this level, the electric and magnetic fields are given by Maxwell's equations, but they could also be traced back to the influence of charged particles. The fields are convenient, but perhaps we could do without them. Part b gives something new. In this contribution to the force, the electromagnetic fields are affected by spontaneous activity (virtual particle -antiparticle pairs) in the electron field.

the action is. But we can "look" inside the computers' calculations, to see what the quarks and gluon fields are up to. To nimbler eyes, space would look like the ultrastroboscopicmicronano lava lamp you see in Color Plate 4. Creatures with such eyes wouldn't share the human illusion that space is empty.

Material Grid

Besides the fluctuating activity of quantum fields, space is filled with several layers of more permanent, substantial stuff. These are ethers in something closer to the original spirit of Aristotle and Descartes—they are materials that fill space. In some cases, we can even identify what they're made of and even produce little samples of it. Physicists usually call these material ethers *condensates.* One could say that they (the ethers, not the physicists) condense spontaneously out of empty space as the morning dew or an all-enveloping mist might condense out of moist, invisible air.

The best understood of these condensates consists of quark-antiquark pairs. Here, we are talking about real particles, beyond the ephemeral, virtual particles that spontaneously come and go. The usual name for this space-filling mist of quarks and antiquarks is *chiral symmetry-breaking condensate*, but let's just call it after what it is: $Q\bar{Q}$ (pronounced "Q-Q bar"), for quark-antiquark.

For $Q\bar{Q}$, as for the other condensates, the two main questions are

- Why do we think it exists?
- How can we verify that it's there?

Only in the case of $Q\bar{Q}$ do we have good answers to both questions.

$Q\bar{Q}$ forms because perfectly empty space is unstable. Suppose we clean out space by removing the condensate of quark-antiquark pairs—something we can do more easily in our minds, with the help of equations and computers, than in laboratory experiments. Then, we compute, quark-antiquark pairs have negative total energy. The mc^2 energy cost of making those particles is more than made up by the energy you can liberate by unleashing the attractive forces between them, as they bind into little molecules. (The proper name for these quark-antiquark molecules is σ mesons.) So perfectly empty space is an explosive environment, ready to burst forth with real quark-antiquark molecules.

Chemical reactions usually start with some ingredients A, B and produce some products C, D; then we write

$$A + B \rightarrow C + D$$

and, if energy is liberated,

$$A + B \rightarrow C + D + \text{energy}$$

(This is the equation for an explosion.)

In that notation, our reaction is

$$[\text{nothing}] \rightarrow \text{quark} + \text{antiquark} + \text{energy}$$

—no starting ingredients (other than empty space) required! Fortunately, the explosion is self-limiting. The pairs repel each other, so as their density increases it gets harder to fit new ones in. The total cost for producing a new pair includes an extra fee, for interacting with the pairs that are already there. When there's no longer a net profit, production stops. We wind up with the space-filling condensate, $Q\bar{Q}$, as the stable endpoint.

An interesting story, I hope you'll agree. How do we know it's right?

One answer is that it is a mathematical consequence of equations—the equations of QCD—that we have many other ways of checking. But although that may be a perfectly *logical* answer (the checks, as we've discussed, are very detailed and convincing), it's not exactly science at its best. We'd like the equations to have consequences we can see reflected in the physical world.

A second answer is that we can calculate the consequences of $Q\bar{Q}$ itself, and check whether they match things we see in the physical world. To be more specific, we can calculate whether $Q\bar{Q}$, considered as a material, can vibrate, and what the vibrations should look like. This is very close to what "luminiferous ether" fans once wanted to have for light: a good old-fashioned material, more substantial than electromagnetic fields. Vibrations of $Q\bar{Q}$ aren't visible light, but they do describe something quite definite and observable, namely π mesons. Among the hadrons, π mesons have unique properties. They are by far the lightest, for example,[7] and they never fit comfortably within the quark model. So it's very satisfying—and after you study the details in depth, it's very convincing—that they arise in quite a different way, as vibrations of $Q\bar{Q}$.

7. For experts: they decouple at low energies, as well.

A third answer is the most direct and dramatic of all, at least in principle. We started by considering the *thought* experiment of cleaning out space. How about doing it for real? Scientists at the relativistic heavy ion collider (RHIC) at Brookhaven National Laboratory, on Long Island, have been working on it, and more such work will be going on at the LHC. What they do is accelerate two big collections of quarks and gluons moving in opposite directions—in the form of heavy atomic nuclei, such as gold or lead nuclei—to very high energy, and make them collide. This is not a good way to study the basic, elementary interactions of quarks and gluons or to look for subtle signs of new physics, because many many such collisions will happen at once. What you get, in fact, is a small but extremely hot fireball. Temperatures over 10^{12} degrees (Kelvin, Celsius, or Fahrenheit—at this level, you can take your pick) have been measured. This is a billion times hotter than the surface of the Sun; temperatures this high last occurred only well within the first second of the big bang. At such temperatures, the condensate $Q\bar{Q}$ vaporizes: the quark-antiquark molecules from which it's made break apart. So a little volume of space, for a short time, gets cleaned out. Then, as the fireball expands and cools, our pair-forming, energy-liberating reaction kicks in, and $Q\bar{Q}$ is restored.

All this almost certainly happens. "Almost" comes in, though, because what we actually get to observe is the flotsam and jetsam thrown off as the fireball cools. Color Plate 5 is a photograph of what it looks like. Obviously, the photograph doesn't come labeled with circles and arrows telling you what's responsible for each aspect of this spectacularly complicated mess. You have to interpret it. In this case, even more (actually, much more) than with the pictures of proton interiors and jets that we discussed in Chapter 6, the interpretation is a complicated business. Today, the most accurate and complete interpretations build in the process of $Q\bar{Q}$ melting and re-formation we've been discussing, but they're not yet as clear and convincing as we might hope for. People continue to work at it—both the experiments and the interpretation.

For the next-best-understood condensate, we have good circumstantial evidence that it exists, but only guesses about what it's made up of. The evidence comes from a part of fundamental physics we haven't mentioned so far, the theory of the so-called weak interaction.[8] We have a good theory of the weak interaction that's gone from triumph to triumph since the early 1970s. Notably, this theory was used to predict the existence, mass, and detailed properties of the *W* and *Z* bosons before they were observed experimentally. The theory is usually called the *standard model*, or the Glashow-Weinberg-Salam model, after Sheldon Glashow, Steven Weinberg, and Abdus Salam, three theorists who played key roles in formulating it (for which they shared the Nobel Prize in 1979).

In the standard model, the *W* and *Z* bosons play leading roles. They satisfy equations very similar to the equations for gluons in the quantum chromodynamics. Both are symmetry-based expansions of the equations for photons in quantum electrodynamics (namely, Maxwell's equations). Activity in the *W* and *Z* boson fields creates the weak interactions, in the same sense that activity in the photon field is responsible for electromagnetism, and activity in the color gluon fields is responsible for the strong interaction.

The striking similarities among our fundamental theories of superficially very different forces hint at the possibility of a synthesis, in which all of them will be seen as different aspects of a more encompassing structure. Their different symmetries might be sub-symmetries of a larger master symmetry. Extra symmetry allows the equations to be rotated into themselves in even more ways; that is, there are more ways of making "distinctions without differences." Thus it opens new possibilities for making connections among patterns that previously seemed unrelated. If our fundamental equations describe partial patterns that we can make more symmetric, by making additions, we're tempted to think

---

8. For more details on the weak interaction, see the glossary, Chapter 17, and Appendix B.

that maybe they *really are* just facets of the larger, unified structure. Anton Chekhov famously advised,

> If in Act One there is a rifle hanging over the mantelpiece, it must have been fired by the fifth act.

Now I've hung the rifle of unification.

Returning to the standard model: The *W* and *Z* bosons are attractive lead players, but they need help to fit the parts they're meant to play. Left to themselves, according to the equations that define them, they'd be massless, like the photon and the color gluons. Reality's script, however, calls for them to be heavy. It's as if Tinkerbelle had been cast as Santa Claus. To enable the sprite to impersonate the plumpkin, we've got to pad her out in a pillowy costume.

Physicists know how to do this trick—that is, to make the *W* and *Z* bosons acquire mass. We think. In fact Nature showed us how, by giving us a demonstration. My wife, who's an accomplished writer and a fountain of good advice, gave me a list of cliché words to avoid, including *amazing, astonishing, beautiful, breathtaking, extraordinary*, and others you can probably guess. I mostly follow her advice. But I have to say that I find what I'm about to tell you amazing, astonishing, beautiful, breathtaking, and, yes, extraordinary.

The model Nature gives us for making force-carrying particles heavy is superconductivity. For inside superconductors, *photons* become heavy! I've offloaded a more detailed discussion of this to Appendix B, but here's the essential idea. Photons, as we've discussed, are moving disturbances in electric and magnetic fields. In a superconductor, electrons respond vigorously to electric and magnetic fields. The electrons' attempt to restore equilibrium is so vigorous that they exert a kind of drag on the fields' motion. Instead of moving at the usual speed of light, therefore, inside a superconductor photons move more slowly. It's as if they've acquired inertia. When you study the equations, you find that the slowed-down photons inside a superconductor obey the same equations of motion as would particles with real mass.

If you happened to be a life-form whose natural habitat was the interior of a superconductor, you'd simply perceive the photon as a massive particle.

Now let's turn the logic around. Humans are a life-form that observes, in its natural habitat, photon-like particles, the *W* and *Z* bosons, that are massive. So perhaps we humans should suspect that we live inside a superconductor. Not, of course, a superconductor in the ordinary sense, that's supergood at conducting the (electric) charge that photons care about, but rather a superconductor for the charges that *W* and *Z* bosons care about. The standard model is based on that idea; and, as we've said, the standard model is very successful at describing reality—the reality we find ourselves inhabiting. Thus we come to suspect that the entity we call empty space is an exotic kind of superconductor.

Where you have superconductivity, there's got to be a material that does the conducting. Our exotic superconductivity works everywhere, so the job requires a space-filling material ether.

Big Question: What is that material, concretely? What is it that, for the cosmic superconductor, plays the role that electrons play in ordinary superconductors?

Unfortunately, it can't be the material ether we understand well, $Q\bar{Q}$. Actually $Q\bar{Q}$ *is* an exotic superconductor of the right kind, and it *does* contribute to the *W* and *Z* boson masses. But it falls short quantitatively by a factor of about a thousand.

No presently known form of matter has the right properties. So we don't really know what this new material ether is. We know its name: the Higgs condensate, after Peter Higgs, a Scots physicist who pioneered some of these ideas. The simplest possibility, at least if you equate simplicity with adding as little as possible, is that it's made from one new particle, the so-called Higgs particle. But the cosmic superconductor could be a mixture of several materials. In fact, as we've mentioned, we already know that $Q\bar{Q}$ is part of the story, though a small part. As we'll see later, there are good reasons to suspect that a whole new world of particles is ripe for discovery, and that several among them chip in to the cosmic superconductor, a.k.a. the Higgs condensate.

Taken at face value, the most promising unified theories[9] seem to predict the existence of all kinds of particles we haven't yet observed. Additional condensates might save the day. New condensates can make the unwanted particles very heavy—just as the Higgs condensate does for the *W* and *Z* bosons, only more so. Particles with very large mass are hard to observe. It takes more energy, and hence bigger accelerators, to produce them as real particles. Even their indirect influence, as virtual particles, is diminished.

Of course, it would be cheap speculation to add new stuff into your equations just because you know how to make excuses when it isn't observed. What makes the unified field theories interesting is that they explain features of the world that we observe, and—better yet—predict new ones. Now I've told you that the rifle is loaded.

The entity we perceive as empty space is a multilayered, multicolored superconductor. What an amazing, astonishing, beautiful, breathtaking concept. Extraordinary, too.

The Mother of All Grids: Metric Field

Here's the Einstein quotation I saved up. In 1920 he wrote,

> According to the general theory of relativity space without ether is unthinkable; for in such space there not only would be no propagation of light, but also no possibility of existence for standards of space and time (measuring-rods and clocks), nor therefore any space-time intervals in the physical sense.

It serves as a suitable introduction to the mother of all Grids: the metric field.

Let's begin with something simple and familiar: maps of the world. Because maps are flat, whereas the thing they're meant to depict—the surface of Earth—is (approximately) spherical,

9. That is, the ones I think are most promising—the ones we'll be discussing in Chapters 17–21.

obviously the maps require interpretation. There are many ways of making a map that represents the geometry of the surface it depicts. All use the same basic strategy. The crucial thing is to lay down a grid of instructions for how to do geometry locally. More specifically, in each little region of the map, you lay down which direction corresponds to north and which direction corresponds to east (south and west will be the opposite directions, of course). You also specify, in each direction, what interval on the map corresponds to a mile—or kilometer, or lightmillisecond, or whatever—on Earth.

For example, maps based on the standard Mercator projection maintain north vertical and east horizontal. Then the surface of Earth can be fit to a rectangle. Going "around the world" from west to east takes you horizontally from one side of the map to the other, whether you follow the equator or the arctic circle. Back on Earth the equator covers a much longer distance than the arctic circle, so taking the map at face value gives a distorted impression: the polar regions appear, relatively, much larger than they are on Earth. But the grid instructs you how to get the distances right. In the polar regions you should use bigger rulers! (Things get a little crazy right at the poles. The whole top line on the map corresponds to a single point on Earth, namely the North Pole, and the whole bottom line corresponds to the South Pole.)

All the information you need to reconstruct the geometry of Earth's surface from the map is contained in the grid of instructions.[10] For example, here's how you can tell the map describes a sphere. First pick a point on the map. Then, for each direction, measure off a fixed distance r from the reference point (following the grid instructions), and make a dot. The dotted places on the map now correspond to all places on Earth that are distance r from

10. Technical point: To measure the length of a path that goes in directions other than the local north-south or east-west, you break the path into little steps, use Pythagoras' theorem on each one, and add up the lengths. The smaller the steps, the more accurate the measurement.

the reference point. Connect the dots. Generally (for instance, if your map is constructed à la Mercator), the figure you get on the map won't look like a circle, even though it represents a circle on Earth. Nevertheless, you can use the map to measure the distance around the circle on Earth that the figure represents. And you'll find it's less than $2\pi r$. (For experts: It will be $R \sin(2\pi r/R)$, where R is the radius of Earth.) If the map depicted a flat surface—which might not be obvious, if you're using a distorted grid—you'd get exactly $2\pi r$. You could also find a circumference greater than $2\pi r$. Then you'd have discovered that your map is describing a saddle-shaped surface. Spheres, naturally, are defined to have positive curvature; flat surfaces have zero curvature; saddles have *negative* curvature.

Although visualization becomes much more difficult, the same ideas apply to three-dimensional space. Instead of a grid of instructions for doing geometry on a sheet, we can consider a grid of instructions that fills out a three-dimensional region. Such bulked-up "maps" contain (as slices) the sort of two-dimensional maps we have just discussed, as well as specs for putting the slices together. They define curved three-dimensional spaces.

So instead of working directly with complicated three-dimensional shapes that are (at best) extremely difficult to visualize, we can work in ordinary space, using grids of instructions. We can work with these maps without having to sacrifice any information.

The grid of instructions for doing geometry locally is called, in the scientific literature, the *metric field*. The lesson of maps is that the geometry of surfaces, or curved spaces of higher dimensions, is equivalent to a grid, or field, of instructions for how to set directions and measure distance locally. The underlying "space" of the map can be a matrix of points, or even an array of registers in a computer. With the proper grid of instructions, or metric field, either of those abstract frameworks can represent complicated geometry faithfully. Map makers and computer graphics wizards are expert at exploiting those possibilities.

We can also add time to the story. Special relativity tells us that one guy's time is another guy's mixture of space and time, so it's natural to treat space and time on the same footing. To do that we need a four-dimensional array. The instruction grid, or metric field, at each point specifies which three directions are to be considered spatial directions—you can call them north, east, and up, although if you're mapping deep space those names are quaint[11]—and the standards of length in those directions. It also specifies that another direction represents time and gives a rule for translating map lengths in that direction into intervals of time.

In the general theory of relativity, Einstein used the concept of curved space-time to construct a theory of gravity. According to Newton's second law of motion, bodies move in a straight line at constant velocity unless a force acts upon them. The general theory of relativity modifies this law to postulate that bodies follow the straightest possible paths through space-time (so-called geodesics). When space-time is curved, even the straightest possible paths acquire bumps and wiggles, because they must adapt to changes in the local geometry. Putting these ideas together, we say that bodies respond to the metric field. These bumps and wiggles in a body's space-time trajectory—in more pedestrian language, changes in its direction and speed—provide, according to general relativity, an alternative and more accurate description of the effects formerly known as gravity.

We can describe general relativity using either of two mathematically equivalent ideas: curved space-time or metric field. Mathematicians, mystics, and specialists in general relativity tend to like the geometric view because of its elegance. Physicists trained in the more empirical tradition of high-energy physics and quantum field theory tend to prefer the field view, because it corresponds better to how we (or our computers) do concrete calculations. More important, as we'll see in a moment, the field

---

11. Mathematicians and physicists usually call them x_1, x_2, x_3—less quaint, more opaque.

view makes Einstein's theory of gravity look more like the other successful theories of fundamental physics, and so makes it easier to work toward a fully integrated, unified description of all the laws. As you can probably tell, I'm a field man.

Once it's expressed in terms of the metric field, general relativity resembles the field theory of electromagnetism. In electromagnetism, electric and magnetic fields bend the trajectories of electrically charged bodies, or bodies containing electric currents. In general relativity, the metric field bends the trajectories of bodies that have energy and momentum. The other fundamental interactions also resemble electromagnetism. In QCD, the trajectories of bodies carrying color charge are bent by color gluon fields; in the weak interaction, still other types of charge and fields are involved; but in all cases the deep structure of the equations is very similar.

These similarities extend further. Electric charges and currents affect the strength of the electric and magnetic fields nearby—that is, their average strength, ignoring quantum fluctuations. This is the "reaction" of fields corresponding to their "action" on charged bodies. Similarly, the strength of the metric field is affected by all bodies that have energy and momentum (as all known forms of matter do). Thus the presence of a body A affects the metric field, which in turn affects the trajectory of another body B. This is how general relativity accounts for the phenomenon formerly known as the gravitational force one body exerts on another. It vindicates Newton's intuitive rejection of action at a distance, even as it dethrones his theory.

Consistency requires the metric field to be a *quantum* field, like all the others. That is, the metric field fluctuates spontaneously. We do not have a satisfactory theory of these fluctuations. We know that the effects of quantum fluctuations in the metric field are usually—in our experience so far, always—small in practice, simply because we get very successful theories by ignoring them! From delicate biochemistry to exotic goings-on at accelerators to the evolution of stars and the early moments of the big bang, we've been able to make precise predictions, and have seen them accurately verified, while ignoring possible quantum fluctuations in the metric field. Moreover, the modern GPS system maps out space and time directly. It doesn't allow for quantum gravity, yet it works very well. Experimenters have worked very hard to discover *any* effect that could be ascribed to quantum fluctuations in the metric field, or, in other words, to quantum gravity. Nobel Prizes and everlasting glory would attend such a discovery. So far, it hasn't happened.[12]

Nevertheless, the Chicago objection—"That works in practice, but what about in theory?"—still holds. The problem that arises is much like the problems we saw with the quark model, and especially the parton model, in Chapter 6. Worrying about those theoretical problems eventually led to the concept of asymptotic freedom and to a complete, extremely successful theory of quarks and the (newly predicted!) color gluons. The analogous problem for quantum gravity hasn't been solved. Superstring theory is a valiant attempt but very much a work in progress. At present it's more a collection of hints about what a theory might look like than a concrete world-model with definite algorithms and predictions. And it hasn't deeply incorporated the basic Grid ideas. (For experts: string field theory is clumsy at best.)

In the quotation that opened this section, Einstein said that space-time without the metric field is "unthinkable." Taken literally, that's obviously false—it's easy to think about it! Let's go back to our map. If the grid instructions are erased or get lost, the map still tells us things. It tells us which countries are next to which, for example. It just wouldn't tell us, reliably, how big they are, or what shape. Even without information about size and shape, we still have what's called topological information. That still leaves plenty to think about.

What Einstein meant is that it's hard to imagine how the physical world would function without the metric field. Light wouldn't

12. But I've advertised a promising opportunity in the endnotes.

know which way to move or how fast; rulers and clocks wouldn't know what they're supposed to measure. The equations Einstein had for light, and for the materials out of which you might make rulers and clocks, can't be formulated without the metric field.

True enough, but a lot of things in modern physics are hard to imagine. We have to let our concepts and equations take us where they will. What Hertz said about this is so important (and so well expressed) that it bears repeating:

> One cannot escape the feeling that these mathematical formulae have an independent existence and an intelligence of their own, that they are wiser than we are, wiser even than their discoverers, that we get more out of them than was originally put into them.

In other words, our equations—and more generally, our concepts—are not just our products but also our teachers.

In that spirit, the discovery that the Grid is filled with several materials, or condensates, raises an obvious question: Is the metric field a condensate? Might it be made from something more fundamental? And that question raises another: Could the metric field, like $Q\bar{Q}$, have vaporized at the origin of the universe, in the earliest moments of the big bang?

A positive answer would open up a new way of addressing a question that vexed Saint Augustine: "What was God doing *before* He created the world?" (Subtext: What was He waiting for? Wouldn't it have been better to start sooner?) Saint Augustine gave two answers.

> First answer: Before God created the world, He was preparing Hell for people who ask foolish questions.
>
> Second answer: Until God creates the world, no "past" exists. So the question doesn't make sense.

His first answer is funnier, but the second, spelled out at length in Chapter 10 of Augustine's *Confessions*, is more interesting. Augustine's basic argument is that the past no longer exists and the future does not yet exist; properly speaking, there is only the present. But the past has a sort of existence within minds, as present memory (as does of course the future, as present expectations). Thus the existence of a past depends on the existence of minds, and there can be no "before" in the absence of minds. Before minds were created, there was no before!

A modern, secular version of Augustine's question is "What happened before the big bang?" And a version of his second answer based on physics might apply. Not that minds are necessary for time—I don't think many physicists would accept that (and the equations of physics certainly don't). But if the metric field vaporizes, with it goes the standard of time. Once no *clocks* exist (and this means an end not just to elaborate time-keeping devices, but to every physical process that could serve to mark time), time itself, along with the whole notion of "before," loses any meaning. The flow of time commences with the condensation of the metric field.

Could the metric field change in some other way (crystallize?) under pressure, for example near the center of black holes? (We know that quarks will form weird condensates under pressure, with funny names like color-flavor locked superconductor, that are different from $Q\bar{Q}$.)

Could the more fundamental material from which the metric field is made be the same material we need to unify the other forces?

Great questions, I hope you'll agree! Unfortunately, we don't have worthy answers yet. (I'm working on it. . . .) But it's a sign of our progress, and of the growth of our ambition, that we can formulate questions and think seriously about possibilities that Einstein considered "unthinkable." We have better equations now, and richer concepts, and we let them be our guides.

Grid Weighs

Mass traditionally has been regarded as *the* defining property of matter, the unique feature that gives substance to substance. So the recent astronomical discovery that Grid weighs—that the entity we perceive as empty space has a universal, nonzero density—crowns the case for its physical reality. Although it's somewhat peripheral to the main thrust of this book, I'll take a few pages to discuss the nature of that discovery and its cosmological implications, because it's both fundamentally important and extremely interesting.[13]

The concept of Grid density is essentially the same as Einstein's cosmological term, which is essentially the same as "dark energy." There are slight differences in interpretation and emphasis, which I'll explain as they come up, but all three terms refer to the same physical phenomenon.

In 1917 Einstein introduced a modification of the equations he originally proposed for general relativity two years earlier. His motivation was cosmology. Einstein thought that the universe had constant density, both in time and (on average) in space, so he wanted to find a solution with those properties. But when he applied his original equations to the universe as a whole, he could not find such a solution. The underlying problem is easy to grasp. It was anticipated by Newton in 1692, in a famous letter to Richard Bentley:

> . . . [I]t seems to me that if the matter of our sun and planets and all the matter of the universe were evenly scattered throughout all the heavens, and every particle had an innate gravity toward all the rest, and the whole space throughout which this matter was scattered was but finite, the matter on the outside of this space would, by its gravity, tend toward all the matter on the inside and, by consequence, fall down into the middle of the whole space and there compose one great spherical mass. But if the matter was evenly disposed throughout an infinite space, it could never converge into one mass; but some of it would converge into one mass and some into another . . .

---

13. To avoid introducing too many complications at once, I've deferred discussion of another extremely interesting astronomical discovery, "dark matter." We'll get to it later.

Simply put: gravity is a universal attraction, so it is not content to leave things separate. Because gravity is always trying to bring things together, it's not terribly surprising that you can't find a solution where the universe maintains a constant density.

To get the kind of solution he wanted, Einstein changed the equations. But he changed them in a very particular way that doesn't spoil their best feature, namely that they describe gravity in a way consistent with special relativity. There is basically only one way to make such a change. Einstein called the new term that he added to the equations for gravity the "cosmological term." He didn't really offer a physical interpretation of it, but modern physics supplies a compelling one, which I'll describe momentarily.

Einstein's motivation for adding the cosmological term, to describe a static universe, soon became obsolete, as evidence for the expansion of the universe firmed up in the 1920s, mainly through the work of Edwin Hubble. Einstein called the ideas that caused him to miss predicting the expansion of the universe his "greatest blunder." (And it really was a blunder, because the model universe he produced, even with his new equations, is unstable. Strictly uniform density is a solution, but any small deviation from uniformity increases with time.) Nevertheless, the possibility he identified, of adding a new term to the equations of general relativity without spoiling the theory, was prophetic.

The cosmological term can be viewed in two ways. Like $E = mc^2$ and $m = E/c^2$, they are equivalent mathematically but suggest different interpretations. One way (the way Einstein viewed it) is as a modification of the law of gravity. Alternatively, the term can be viewed as the result of having a constant density of mass and also a constant pressure everywhere in space and for all time. Because both mass-density and pressure have the same value everywhere, they can be regarded as intrinsic properties of space itself. That's the Grid viewpoint. If we take it as given that space has these properties, and focus exclusively on the gravitational consequences, we arrive back at Einstein's viewpoint.

A key relationship governing the physics of the cosmological term relates its density ρ to the pressure p it exerts, using the speed of light c. There's no standard name for this equation, but it will be handy to have one. I'll call it the well-tempered equation, because it prescribes the proper way to tune a Grid. The well-tempered equation reads

$$\rho = -p/c^2 \qquad (1)$$

Where does it come from? What does it mean?

The well-tempered equation looks like a mutated clone of Einstein's second law, $m = E/c^2$. The m has turned into a ρ, and the E into a p—and there's that – sign—but you can't help noticing a resemblance. And in fact they are deeply related.

Einstein's second law relates the energy of an isolated body at rest to its mass. (See Chapter 3 and Appendix A.) It is a consequence of special relativity theory, though not an immediately obvious one. In fact it did not appear in Einstein's first relativity paper; he wrote a separate note about it afterwards.

The well-tempered equation is likewise a consequence of special relativity, but now applied to a homogeneous, space-filling entity rather than to an isolated body. It's not immediately obvious how a nonzero Grid density can be consistent with special relativity. To appreciate the problem, think about the famous Fitzgerald-Lorentz contraction, which we encountered in Chapter 6. To an observer moving at constant velocity, objects appear foreshortened in the direction of motion. It would seem, therefore, that the moving observer would see a higher Grid density. That's contrary to relativity's boost symmetry, which says she must see the same physical laws.

The pressure that goes with density, according to the well-tempered equation, provides a loophole. The weighing scales of the moving observer, according to the equations of special relativity, register a new density that is a mixture of the old density and the old pressure—just as, perhaps more familiarly, her clocks register time intervals that are mixtures of the old time intervals and the old space intervals. If—and only if—the old density and the old pressure are related in just the way prescribed by the well-tempered equation, then the new density (and the new pressure) will be the same as the old.

Another, closely related consequence of the well-tempered equation is central to the cosmology of Grid density. In an expanding universe, the density of any normal kind of matter will go down. But the density of the well-tempered grid stays constant! If you're up for a little exercise in first-year physics and algebra, here comes a pretty connection tying that constancy of density directly to Einstein's second law. (If not, just skip the next paragraph.)

Consider a volume V of space, filled with Grid density ρ. Let the volume expand by δV. Ordinarily, as a body expands under pressure it does work, and so loses energy. But the $-$ sign in the equation for a well-tempered Grid gives us *negative* pressure $p = -\rho c^2$. So by expanding, our well-tempered Grid *gains* energy $\delta V \times \rho c^2$. According to Einstein's second law, therefore, its mass increases by $\delta V \times \rho$. And that's just enough to fill the added volume δV with density ρ, allowing the density of the Grid to remain constant.

Each of the Grid components we've discussed—fluctuating quantum fields of many sorts, $Q\bar{Q}$, Higgs condensate, unification-salvaging condensate, space-time metric field (or condensate?)—is well-tempered. Each of these space-filling entities obeys the well-tempered equation, because each is consistent with the boost symmetry of special relativity.

It's possible to measure the cosmic density and the pressure separately, using quite different techniques. The density affects the curvature of space, which astronomers can measure by studying the distortion such curvature causes in images of distant galaxies, or—a powerful new technique—in the cosmic microwave background radiation. Using the new technique, by 2001 several groups were able to prove that there was much more mass in the universe than could be accounted for by normal matter alone.

About 70% of the total mass appears to be very uniformly distributed, both in space and time.

The pressure affects the rate at which the universe is expanding. That rate can be measured by studying distant supernovae. Their brightness tells you how far away they are, while the redshift of their spectral lines tells you how fast they're moving away. Because the speed of light is finite, when we observe the farther-away ones we're looking at their past. So we can use supernovae to reconstruct the history of expansion. In 1998 two powerhouse teams of observers reported that the rate of expansion of the universe is increasing. This was a big surprise, because ordinary gravitational attraction tends to brake the expansion. Some new effect was showing up. The simplest possibility is a universal negative pressure, which encourages expansion.

The term *dark energy* became a shorthand for both of these discoveries: the additional mass and the accelerating expansion. It was meant to be agnostic about the relative values of density and pressure. If we simply called both of them the cosmological term, we'd be prejudging their relative magnitudes. But apparently we'd be right. The two very different quantities, cosmic mass density and cosmic pressure, observed in very different ways, do seem to be related by $\rho = -p/c^2$.

Is the astronomical discovery that space weighs, and seems to obey the well-tempered equation, a brilliant confirmation of the deep structures upon which we erect our best models of the world? Yes and no. To be honest, probably I should write yes and NO.

The problem is that the total density that astronomers weigh is far, far smaller than simple estimates of what any of our condensates provides. Here are simple estimates of the densities involved, as multiples of what the astronomers actually find:

- Quark-antiquark condensate: 10^{44}
- Weak superconducting condensate: 10^{56}
- Unified superconducting condensate: 10^{112}

- Quantum fluctuations, without supersymmetry: ∞
- Quantum fluctuations, with supersymmetry:[14] 10^{60}
- Space-time metric: ? (The physics here is too murky to allow a simple estimate.)

If any of these simple estimates were correct, the evolution of the universe would be *much* more rapid than what's observed.

Why is the real density of space much smaller? Maybe there's a vast conspiracy among these and possibly other contributions, some necessarily negative, to give a total that's very much smaller than each individual contribution. Maybe there's an important gap in our understanding of how gravity responds to Grid density. Maybe both. We don't know.

Before dark energy was discovered, most theoretical physicists, looking at the enormous discrepancy between simple estimates of the density of space and reality, hoped that some brilliant insight would supply a good reason why the true answer is zero. Feynman's "*because it's empty*" was the best, or at least the most entertaining, idea I heard along those lines. If the answer really isn't zero, we need different ideas. (It's still logically possible that the ultimate density is zero, and that the universe is very slowly settling toward that value.)

A popular speculation today is that many different possible condensates contribute to the density, some positive, some negative. It's only when the contributions cancel almost completely that you get a nice slowly evolving universe that is sufficiently user-friendly to be observed. An observable universe has to allow enough quality time for potential observers to evolve. Thus (according to this speculation) we observe an improbably small total Grid density because if the total were much larger, nobody would be around to observe it. Maybe that's right, but it's a diffi-

---

14. We'll be discussing supersymmetry in more depth later, in connection with unification. The main thing to note here is that it suggests a ridiculously large contribution to the density, just like everything else.

cult idea to make precise, or to check. Sometimes we can leverage uncertainty into precision by gathering many samples. We do that when making insurance tables or applying quantum mechanics. But for the universe, we're stuck with a sample size of one.

Anyway, in the one universe we've had a look at, Grid weighs. Fortunately, to clinch *that* conclusion, one universe is enough.

Recapitulation

At the beginning of this chapter, I advertised key properties of the Grid, that ur-stuff that underlies physical reality:

- The Grid fills space and time.
- Every fragment of Grid—each space-time element—has the same basic properties as every other fragment.
- The Grid is alive with quantum activity. *Quantum* activity has special characteristics. It is spontaneous and unpredictable. And to observe quantum activity, you must disturb it.
- The Grid also contains enduring, material components. The cosmos is a multilayered, multicolored superconductor.
- The Grid contains a metric field that gives space-time rigidity and causes gravity.
- The Grid weighs, with a universal density.

Now, after the sales pitch, I hope you buy it!

9

Computing Matter

A play of Bits outputs our Its.

JOHN WHEELER HAD A GIFT for capturing profound ideas in catchy phrases. "Black hole" is probably his best-known creation, but my favorite is "Its From Bits." These three words capture an inspiring ideal for theoretical science. We try to find mathematical structures that mirror reality so completely that no meaningful aspect escapes them. Solving the equations tells us both what exists and how it behaves. By achieving such a correspondence, we put reality in a form we can manipulate with our minds.

Philosophical realists claim that matter is primary, brains (minds) are made from matter, and concepts emerge from brains. Idealists claim that concepts are primary, minds are conceptual machines, and conceptual machines create matter. "Its From Bits" says we do not have to choose between these alternatives. Both can be right at the same time. They describe the same thing using different languages.

The ultimate challenge to "Its From Bits" is to find mathematical structures that mirror conscious experience and flexible intelligence—in a word, thinking computers. That has not been achieved, and people still argue about whether it's possible.[1]

1. Of course it is.

The most impressive example of "Its From Bits" that *has* been achieved is the one I'll describe in this chapter. The algorithms of QCD empower us to program computers to churn out protons, neutrons, and the whole motley crew of strongly interacting particles. Its from bits, indeed!

As a bonus, we achieve another Wheelerism: "Mass Without Mass." The building blocks of protons and neutrons, as revealed by the kinds of experiments we discussed in Chapter 6, are stictly massless gluons and very nearly massless quarks. (The relevant quarks, *u* and *d*, weigh about 1% as much as the protons they make.)

At Brookhaven National Laboratory, on Long Island, and at several other centers around the world, there are special rooms where people rarely tread. Nothing much seems to be happening in these rooms, there's no visible motion, and the only sound is the gentle whir of fans that keep the temperature steady and the humidity low. In these rooms, roughly 10^{30} protons and neutrons are at work. They have been organized into hundreds of computers, harnessed to work in parallel. The team races at teraflop rates, which means 10^{12}—a million million—FLoating point OPerations per second. We let them labor for months—10^7 seconds. At the end, they've done what a single proton does every 10^{-24} second, which is figure out how to orchestrate quark and gluon fields in the best possible way so that they keep the Grid satisfied and make a stable equilibrium.

Why is it so hard?

The Grid is a harsh mistress.

To be more accurate, she's complicated. She has many moods, and she's stormy.

Quantum mechanics works with wave functions that represent many possible configurations of the fields at once, but our classical computers can handle only one configuration at a time. To mimic interactions among many configurations that, in the quantum description, are present simultaneously, a classical computer must:

1. churn for a long time, to produce the configurations

2. store them
3. cross-correlate its ancient memories with its current content

Altogether, there's a poor match between the end and the means. If and when quantum computers become available, we might be in better shape. What's more, the things we're trying to calculate—the particles we observe—constitute small ripples in a turbulent sea of fluctuating Grid. To find the particles, numerically, we have to model the whole sea, and then hunt out the tiny disturbances.

A Toy Model in Thirty-Two Dimensions

When I was but a wee lad I liked to put together, and take apart, plastic model rockets. These models couldn't put up satellites, let alone ferry anyone to the Moon. But they were things I could hold in my hands and play with, and they were aids to imagination. They were built to scale, and there was also a little plastic man on the same scale, so I got a sense of the sizes involved, the difference between an interceptor and a launch vehicle, and some key concepts like payloads and detachable stages. Toy models can be fun and useful.

Similarly, in trying to understand complicated concepts or equations, it's good to have toy models. A good toy model captures some sense of the real thing but is small enough that we can wrap our minds around it.

In the next few paragraphs I'll show you a toy model of quantum reality. It's a vastly simplified model, but I think it's just intricate enough to suggest the vastness of quantum reality. The main point is that quantum reality is REALLY, REALLY BIG.[2] We'll build up a toy model that describes social life among the spins of

2. If you're willing to take my word for it, and would rather avoid the dizzying details, you can proceed directly to the section "The Big (Number) Crunch."

just five particles, and we'll discover that it fills out a space of thirty-two dimensions.

Start with one quantum particle that has a minimal unit of spin. We abstract away—that is, ignore—all its other properties. The resulting object is what is called a quantum bit, or qubit. (For sophisticates: A cold electron trapped in a definite spatial state, say by appropriate electric fields, is effectively a qubit.) The spin of a qubit can point in different directions. We'll write

$$|\uparrow\rangle$$

for the state in which the spin of the qubit is definitely up, and

$$|\downarrow\rangle$$

for the state in which the spin is definitely down.

The qubit can also be in states where the spin points sideways, and that's where the fun begins. It's exactly here, at this juncture, that the central weirdness of quantum mechanics comes into play.

The sideways-pointing states are *not* new, independent states. These sideways-pointers, and all other states of the qubit, are combinations of the states $|\uparrow\rangle$ and $|\downarrow\rangle$ that we already have.

Specifically, for instance, the east-spinning state is

$$|\rightarrow\rangle = \frac{1}{\sqrt{2}}|\uparrow\rangle + \frac{1}{\sqrt{2}}|\downarrow\rangle$$

The state where the spin definitely points east is an equal mixture of north and south. If you measure the spin in the horizontal direction, you'll always find that it points east. But if you measure the spin in the vertical direction, you're equally likely to find that it points north or south. That's the meaning of this strange equation. In more detail, the rule for computing the probability of finding a given result (up or down) when you measure the spin in the vertical direction is that you take the square of the number that multiplies the state with that result. Here, for example, the number $1/\sqrt{2}$ multiplies the spin up state, so the probability of finding spin up is $(1/\sqrt{2})^2 = 1/2$.

This example illustrates, in miniature form, the ingredients that enter into the description of a physical system according to quantum theory. The state of the system is specified by its wave function. You've just seen the wave functions for three specific states. The wave function consists of a set of numbers multiplying each possible configuration of the object being described. (That number might be zero, so if we were being fussy we'd write $|\uparrow\rangle = 1\,|\uparrow\rangle + 0\,|\downarrow\rangle$.) The number multiplying a configuration is called the *probability amplitude* for that configuration. The *square* of the probability amplitude is the probability of observing that configuration.

What about the west-spinning state? By symmetry, it should also have equal probabilities for spin up and for spin down. But it has to be different from the east-spinner. Here it comes:

$$|\leftarrow\rangle = \frac{1}{\sqrt{2}}\,|\uparrow\rangle - \frac{1}{\sqrt{2}}\,|\downarrow\rangle$$

The extra minus sign doesn't affect the probability, because we square it. For east versus west, the probabilities are the same, but the probability amplitudes are different. (In a moment we'll see how the minus sign does have consequences, when we consider several spins at once.)

Now let's consider two qubits. To get the state where both are east-spinners, we multiply two copies of the east-spinning state, and find

$$|\rightarrow\rightarrow\rangle = \frac{1}{2}|\uparrow\uparrow\rangle + \frac{1}{2}|\uparrow\downarrow\rangle + \frac{1}{2}|\downarrow\uparrow\rangle + \frac{1}{2}|\downarrow\downarrow\rangle$$

The probability for finding both spins up is $(1/2)^2 = 1/4$, as is the probability for finding the first up, the second down, and so forth. Similarly, when both are west-spinners, we get

$$|\leftarrow\leftarrow\rangle = \frac{1}{2}|\uparrow\uparrow\rangle - \frac{1}{2}|\uparrow\downarrow\rangle - \frac{1}{2}|\downarrow\uparrow\rangle + \frac{1}{2}|\downarrow\downarrow\rangle$$

Again, all the up-and-down probabilities are equal.

Already, with just these two qubits, we find some behavior that's truly gnarly. (The technical term is *entangled.*) Let's consider two states that we can get by combining the double east-pointer with the double west-pointer.

$$\frac{1}{\sqrt{2}}|\rightarrow\rightarrow\rangle + \frac{1}{\sqrt{2}}|\leftarrow\leftarrow\rangle = \frac{1}{\sqrt{2}}|\uparrow\uparrow\rangle + \frac{1}{\sqrt{2}}|\downarrow\downarrow\rangle$$

$$\frac{1}{\sqrt{2}}|\rightarrow\rightarrow\rangle - \frac{1}{\sqrt{2}}|\leftarrow\leftarrow\rangle = \frac{1}{\sqrt{2}}|\uparrow\downarrow\rangle + \frac{1}{\sqrt{2}}|\downarrow\uparrow\rangle$$

In each of these states, the message of the left-hand expressions is that if we measure the spins in the *horizontal* direction, we'll find either that both are pointing east or that both are pointing west. Each of those possibilities occurs with probability 1/2. We'll never find that one is pointing east and the other west. So as far as measurements in the horizontal direction are concerned, these two states look just the same. It's like knowing you have a matched pair of socks, either black or white, but not knowing which color. That's the message of the left-hand sides of these equations.

The right-hand sides tell you what happens if you measure, in these same states, both spins in the vertical direction. Now the results are very different. In the first state, both spins will be pointing up, or both down; each possibility occurs with probability 1/2. The second state previously (in the last paragraph!) looked the same as the first. Now, viewed from another perspective, it couldn't be more different. In the second state you *never* find the spins pointing in the same vertical direction; if one is up, the other is down.

Either of these states would annoy Einstein, Podolsky, and Rosen, because they exhibit the essence of the famous EPR paradox. Measuring the spin of the first qubit tells you about the result you'll get by measuring the second bit, even though they might be physically separated by a large distance. On the face of it, this "spooky action-at-a-distance," to use Einstein's phrase, seems capable of transmitting information (telling the second spin which way it must point) faster than the speed of light. But that's an illusion, because to get two qubits into a definite state we had to start with them close together. Later we can take them far apart, but if the qubits can't travel faster than the speed of light, neither can any message they carry with them.

More generally, to construct all possible states of two qubits, we add the four possibilities $|\uparrow\uparrow\rangle$, $|\uparrow\downarrow\rangle$, $|\downarrow\uparrow\rangle$, $|\downarrow\downarrow\rangle$, each multiplied by a separate number. That defines a four-dimensional space—you can step off distances in four different directions.[3]

To describe the possible states of five qubits, we have up-or-down choices for each of them (for example, $|\uparrow\downarrow\uparrow\uparrow\downarrow\rangle$ or $|\uparrow\uparrow\downarrow\downarrow\uparrow\rangle$). There are $2 \times 2 \times 2 \times 2 \times 2 = 32$ possibilities, and a general state can contain contributions from all of them, each multiplied by a number. That's how we find ourselves with a thirty-two-dimensional toy model on our hands. Some toy!

Laplace's Demon versus Grid Pandemonium

Pierre-Simon Laplace's masterpiece, the five-volume *Mécanique Céleste,* appeared in installments from 1799 to 1825. It took mathematical astronomy, based on Newtonian principles, to a new level of elegance and precision. Laplace was so impressed with how accurately he was able to calculate celestial motions that he fantasized about what a perfectly knowledgeable calculating

3. To avoid possible confusion: in this way of counting, north and south count as just one direction—stepping 1 mile south is the same as stepping minus 1 mile north.

demon could do. He decided that his demon would be able to predict the future, or reconstruct the past, by calculation:

> An intelligence which, for a given instant, could know all the forces by which nature is animated, and the respective situation of the beings who compose it, if, moreover, it was sufficiently vast to submit these data to analysis, if it could embrace in the same formula the movements of the greatest bodies in the universe as well as those of the lightest atom—nothing would be uncertain for it, and the future, like the past, would be present to its eyes.

Laplace, of course, had in mind a universe based on Newtonian mechanics. How realistic does his demon appear today? Could complete knowledge of the present and limitless mathematical skill reduce the past and the future to calculation?

Grid pandemonium overwhelms Laplace's demon.

Let's first consider the problem the demon is up against. Laplace thought that if you specified the position and velocity of every atom in the world, you specified the world. There would be nothing else to know. And he thought that physics supplied equations relating the complete set of positions and velocities at one time to those at later (or earlier) times. Thus if you knew the state of the world at some time t_0, you could calculate the state of the world at any other time t_1.

With modern quantum theory, the world has become a much bigger place than Laplace could have imagined. Our toy model featured just a handful of qubits[4] but encompassed a thirty-two-dimensional world. The quantum Grid, which embodies our deepest understanding of reality, requires *many qubits at each point in space and time.* The qubits at a point describe the various things that might be happening at that point. For example, one of them describes the probability that (if you look) you will observe an electron with spin up or down, another the probability that (if you look) you will observe an antielectron with spin up or down,

4. One for each finger.

another the probability that (if you look) you will observe a red *u* quark with spin up or down, ... Others describe possible results if you look for photons, gluons, or other particles. On top of that, if space and time are continuous—as the existing laws of physics, so far very successfully, assume—then the number of space-time points is highly infinite.

The world is no longer founded on atoms in the void, so the state of the world no longer consists of the positions and velocities of a lot of atoms. Instead, the world is the tremendous multiple infinity of qubits just described. And to describe its state we must assign a number—a probability amplitude—to every possible configuration of the qubits. In our five-qubit toy model we found that the possible states filled out a space of thirty-two dimensions. The space we must use to describe the state of the Grid, which is our world, brings in infinities of infinities.

A googol is 10^{100}—that is, 1 followed by one hundred zeros. It was meant to be a crazily large number. A googol is, for example, much larger than the number of atoms in the visible universe. But even if we replace all of space by a lattice with just ten points in each direction, and put just one qubit at each point, the *dimension* of the quantum-mechanical version of that schematic model world is much more than a googol. In fact, it's a space whose dimension is more than a googol of googols.

Thus the first part of the demon's task, knowing "the respective situation of the beings who compose" the world, is a tough one. To know the state of the world, he's got to locate a specific point somewhere within a REALLY REALLY BIG space. Compared to this challenge, finding a needle in a haystack is dead easy.

It gets worse. We've talked before about the spontaneous activity of the Grid. It's full of quantum fluctuations, or virtual particles. Those are rough, informal descriptions of a reality we now have the language to express more precisely. To say that the Grid contains spontaneous activity is to say that its state is not a simple one. If we look with high resolution in space and time to see what's going on in the entity we call empty space (as experimenters did at LEP, for example), we find many possible results. Each time we look we see something different. Each observation uncovers a piece of the wave function that describes a typical, very small region of space. Each observation embodies a possibility that occurs, multiplied by some probability amplitude, within that wave function.

So we're looking for a needle that isn't near the bottom of the haystack or in any particularly simple place. It's off to the side, or rather off to this side and that side and the other side and so on in various amounts for googols of googols of sides.

Laplace's imaginary demon is blessed with perfect knowledge of the state of the world. He knows where that needle is. But he's imaginary. Those of us who aren't blessed with perfect knowledge of the state of the world, but would still like to predict something about the future, face some issues. How can we acquire some of the relevant knowledge? How much impact will gaps in our knowledge have?

As Yogi Berra apparently learned from Niels Bohr, "prediction is very difficult, especially about the future." There are (at least) two fundamental reasons why it can be very difficult to predict the future, even if we have all the right equations. One is chaos theory. Roughly speaking, chaos theory says that small uncertainties in your knowledge of the state of the world at time t_0 introduce very large uncertainties in what you can deduce about the state of the world at a significantly later time t_1.

The other is quantum theory. As we've discussed, quantum theory generally predicts probabilities, not certainties. Actually, quantum theory gives you perfectly definite equations for how the wave function of a system changes with time. But when you use the wave function to predict what you'll observe, what it gives you is a set of probabilities for different outcomes.

In view of all this, we've become more humble since Laplace's day about what we can compute, in principle. In practice, however, we're answering questions that Laplace could not begin to imagine, through means he couldn't dream of. For example . . .

The Big (Number) Crunch

Well-informed, modern calculating demons know that they can't simply calculate everything, *à la* Laplace's demon. Their art is to discover aspects of reality that will yield to their craft. Fortunately, chance, uncertainty, and chaos do not infest every aspect of the natural world. Many of the things we're most interested in calculating, such as the shape of a molecule we might use as a drug, the strength of a material we might use to make aircraft, and the mass of a proton, are stable features of reality. Moreover, these systems can be considered in isolation; their properties don't much depend on the state of the world as a whole.[5] In the art of demon calculators, stable, isolated systems are the natural subjects for detailed portraits.

So: fully aware of the difficulties but undaunted, heroes of physics gird their loins, apply for grants, buy clusters of computers, solder, program, debug, even think—whatever it takes to wrest answers from Grid pandemonium.

How do we compute the portrait of a proton?

First, we must replace continuous space and time by a finite structure—a lattice of points—that a computer can handle. That is an approximation, of course, but if the distance between points is small enough, the errors will be small. Second, we must squeeze the REALLY REALLY BIG quantum reality into a classical computing machine. The quantum-mechanical state of Grid lives in a huge space, where its wave function encompasses multitudes of possible patterns of activity. But the computer can manipulate only a few patterns at a time. Because the equations for the evolution in time of any one pattern of activity bring in all the other patterns, the classical computer has to store a vast library of patterns, with their probability amplitudes, in memory. To evolve a current pattern forward in time, it fetches the relevant information about old patterns step by halting step. For each stored pattern, it computes

5. At least, that's a good working hypothesis, and it's justified by its success.

the changes wrought. Finally it stores the evolved probability amplitude for the current pattern, commences to evolve the next pattern, and cycles repeatedly. The Grid is a harsh mistress.

Our eyes were not evolved to resolve distances of order 10^{-14} centimeter, nor our brains to perceive times of order 10^{-24} second. Those capabilities would not help us to avoid predators or find desirable mates. But as our computers cycle through Grid configurations, they are constructing the patterns our eyes would see, were they adapted to those tiny distances and times. Using our noodles, we can improve our vision. That's what gave us Color Plate 4.

Once we've got "empty" space humming, we can pluck it. That is, we can disturb Grid by injecting some extra activity and letting things settle down. If we find stable, localized concentrations of energy, we've found—that is, computed—stable particles. We can match them (if the theory's right!) to protons p, neutrons n, and the rest. If we find localized concentrations of energy that persist for a good while before dissipating, we've found unstable particles. They should match the ρ meson, the Δ baryon, and their kin.

To see what it looks like, examine Color Plate 6 and its caption. That's our deepest understanding of what p, n, ρ, Δ, . . . *are.*

Figure 9.1 shows the very concrete challenge we're up against. It's part of the spectrum of hadrons, strongly interacting particles that have been observed. They have key identifying properties: their mass and spin. The caption provides a technical description of exactly what's been plotted. These details (and there are lots more!) are intricate and full of meaning to experts, but the take-home message is simply that there are plenty of juicy facts for the theory to explain.

Figure 9.2 shows how three of the measured masses are used to fix the parameters of the theory. That is, before doing the calculation, we don't know what masses we should assign to the quarks, or the overall coupling constant. The most accurate way to determine those values is the calculation itself. So we try different values and settle on the ones that give the best fit to the observations.

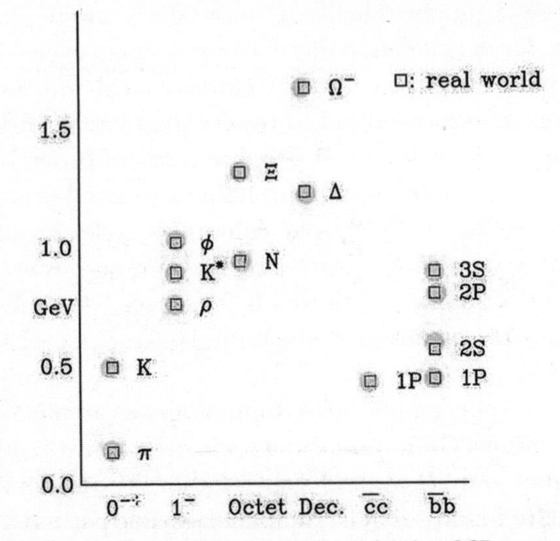

Figure 9.1 A census of strongly interacting particles that QCD must account for. Each point encodes an observed particle. The height of the point indicates the mass of the particle. The first two columns are mesons with spin 0: π, K, and spin 1: ρ, K*, ϕ. The third and fourth columns are baryons with spin 1/2: N, Ξ, and spin 3/2: Δ, Ω, respectively. The fifth and sixth columns are "charmonium" and "bottomonium" mesons with various spins. These mesons are interpreted as bound states of a heavy c (charm) quark and its antiquark, or, respectively, a b (bottom) quark and its antiquark. In these columns the heights represent the mass differences between the particle in question and the lightest possible charmonium or bottomonium state.

If a theory has a lot of parameters, you adjust their values to fit a lot of data, and your theory is not really predicting those things, just accommodating them. Scientists use words like *curve fitting* and *fudge factors* to describe that sort of activity. Those phrases aren't meant to be flattering. On the other hand, if a theory has just a few parameters but applies to a lot of data, it has real power. You can use a small subset of the measurements to fix the parameters; then all other measurements are uniquely predicted.

In that objective sense, QCD is a very powerful theory indeed. Not only doesn't it require many parameters, it doesn't *allow* many: just a mass for each kind of quark, and one universal coupling strength. Furthermore, most of the quark masses are irrelevant for

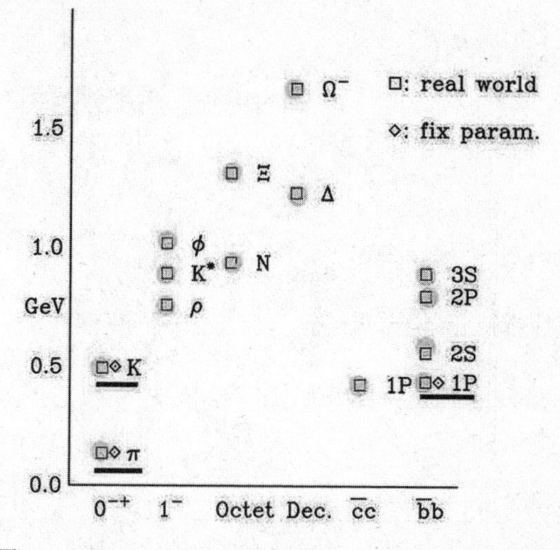

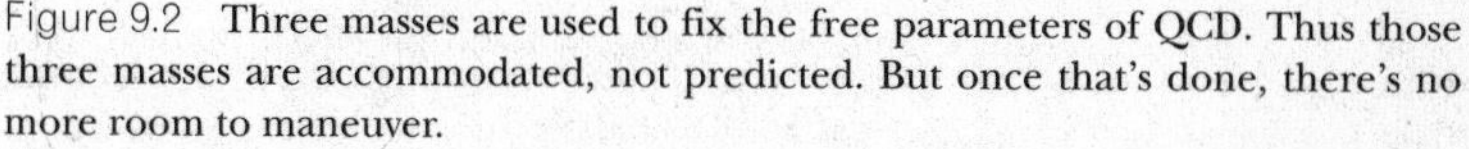

Figure 9.2 Three masses are used to fix the free parameters of QCD. Thus those three masses are accommodated, not predicted. But once that's done, there's no more room to maneuver.

calculating the particle masses in the figure with the precision we can attain; other effects introduce larger uncertainties. We need only the average mass m_{light} of the lightest *u* and *d* quarks and the mass m_s of the strange quark, together with the coupling strength. Having fixed those three things, we have no more room to maneuver. No fudge factors, no excuses, nowhere to hide. If the theory is right, the calculation will match reality. If the calculation doesn't match reality, the theory is irreparably wrong.

Figure 9.3 shows how the calculated values of mass and spin—the unambiguous predictions of QCD—compare with observed values. Because spin comes in discrete units, either the agreement is exact or it's disagreement. So we'd better find observed particles with the exactly the spins and approximately the masses of the predicted particles, and no others. With a sigh of relief, we note that near each "real world" square there's either a "calculated" circle or a "fix parameter" diamond. You see that the calculated

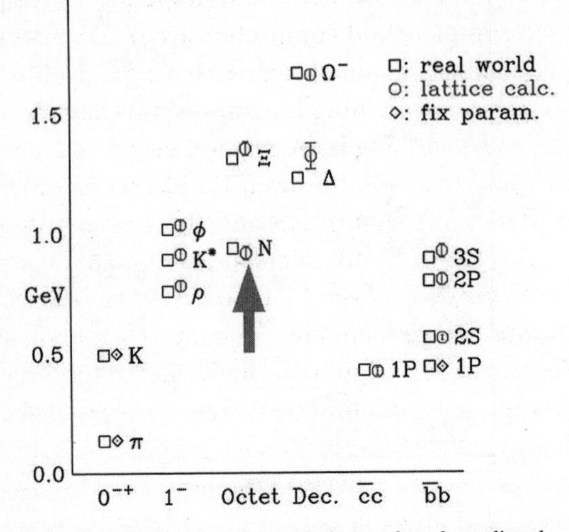

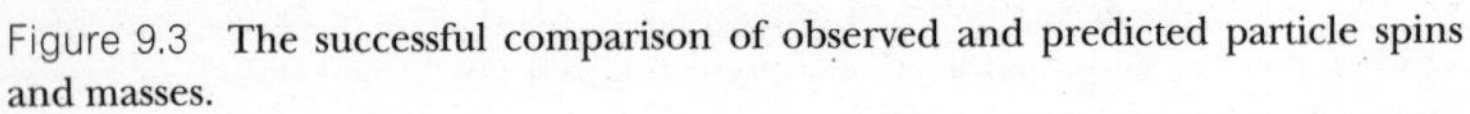

Figure 9.3 The successful comparison of observed and predicted particle spins and masses.

masses agree quite well with the observed values. You'll notice the vertical "error bars" around the calculated values. These reflect the residual uncertainties in the calculation. Various approximations and compromises had to be made because the available computer power, though fantastically large, was finite.

A highlight in this figure is the point labeled N. N stands for nucleon—that is, proton or neutron. (On the scale of this figure, their masses are indistinguishable). QCD succeeds in accounting for the mass of protons and neutrons from first principles. The mass in protons and neutrons, in turn, accounts for overwhelmingly most of the mass of normal matter. I promised to account for the origin of 95% of that mass. There it is.

Also remarkable is what you *don't* see coming out of the computer. There are no extra circles floating around, indicating predicted particles that aren't observed. Especially noteworthy: although the basic inputs to the calculation are quarks and gluons, they don't appear among the outputs! The Principle of Confinement, which seemed so weird and desperate, here appears as a footnote to complete and comprehensive reality-matching.

Of course computing something—or having a gigantic ultra-fast computer compute something for you—is not the same as understanding it. Understanding is the mission of the next chapter.

Before ending this one, however, I'd like to pause for a brief tribute to the austere Figure 9.3 and the community that produced it. Through difficult calculations of merciless precision that call upon the full power of modern computer technology, they've shown that unbendable equations of high symmetry account convincingly and in quantitative detail for the existence of protons and neutrons, and for their properties. They've demonstrated the origin of the proton's mass, and thereby the lioness's share of *our* mass. I believe this is one of the greatest scientific achievements of all time.

10

The Origin of Mass

Knowing how to calculate something is not the same as understanding it. Having a computer calculate the origin of mass for us may be convincing, but it is not satisfying. Fortunately, we can understand it too.

HAVING A COMPUTER SPIT OUT ANSWERS after gigantic and totally opaque calculation does not satisfy our hunger for understanding. What would?

Paul Dirac was famously taciturn, but when he spoke, what he said was often profound. He once said, "I feel I understand an equation, when I can anticipate the behavior of its solutions without actually solving it."

What's the value of such understanding?

"Solving" equations is just one tool—an imperfect one—for working with them. The calculations we discussed in the preceding chapter are an instructive example. They show conclusively that the equations for quark and gluon Grid accurately account for the masses of protons, neutrons, and other hadrons. They also show that those equations keep quarks and gluons hidden. (You can interpret the nonappearance of isolated quarks or gluons as a calculation of *their* mass, when you include their virtual particle clouds—the answer is infinity!)

Those are glorious results, won after heroic efforts by human and machine. But the need for heroic efforts is one of the biggest drawbacks of "solving" equations. We don't want to tie up expensive computer resources and wait a long time for answers every time we ask a slightly different question. Even more important, we don't want to tie up expensive computer resources and wait a *very* long time when we ask more complicated questions. For example, we'd like to be able to predict the masses not only of single protons and neutrons but also of systems containing several protons and neutrons—atomic nuclei. In principle we have the equations to do it, but solving them is impractical. For that matter, we have equations adequate to answer any question of chemistry, *in principle*. But that hasn't put chemists out of business or replaced them with computers, because in practice the calculations are too hard.

In both nuclear physics and chemistry, we are happy to sacrifice extreme precision for ease of use and flexibility. Rather than brutally "solving" the equations by crunching numbers, we make simplified models and find rules of thumb that can give us practical guidance in complicated situations. These models and rules of thumb can grow out of experience in solving the equations, and they can be checked by solving the equations when that's practical, but they have a life of their own. This reminds me of the distinction between graduate students and professors: a graduate student knows everything about nothing, a professor knows nothing about everything. Solving the equations is what graduate students do, understanding them is what professors do.

We are as far as we can be from understanding, when solving the equations reveals behavior that's totally unexpected and appears miraculous. The computers have given us mass—and not just any mass, but *our* mass, the mass of the protons and neutrons we're made from—from quarks and gluons that are themselves massless (or nearly so). The equations of QCD output Mass Without Mass. It sounds suspiciously like something for nothing. How did it happen?

Fortunately, it's possible get a rough, professor-like *understanding* of that apparent miracle. We just have to put together three ideas we've already discussed separately. Let's briefly recall and assemble them.

First Idea: Blossoming Storms

The color charge of a quark creates a disturbance in the Grid—specifically, in the gluon fields—that grows with distance. It's like a strange storm cloud that blossoms from a wispy center into an ominous thunderhead. Disturbing the fields means putting them into a state of higher energy. If you keep disturbing the fields over an infinite volume, the energy cost will be infinite. Even Exxon Mobil wouldn't claim that Nature has the resources to pay that price.[1] So isolated quarks can't exist.

Second Idea: Costly Cancellations

The blossoming storm can be short-circuited by bringing an antiquark, with the opposite color charge, close to the quark. Then the two sources of disturbance cancel, and calm is restored.

If the antiquark were located accurately right on top of the quark, the cancellation would be complete. That would produce the minimal possible disturbance in the gluon fields—namely, none. But there's another price to be paid for that accurate cancellation. It comes from the quantum-mechanical nature of quarks and antiquarks.

According the Heisenberg uncertainty principle, in order to have accurate knowledge of the position of a particle, you must let that particle have a wide spread in momentum. In particular, you must allow that the particle may have *large* momentum. But large momentum means large energy. And the more accurately you fix the position of the particle ("localize" it, in the jargon), the more energy it costs.

1. Maybe I'm being naïve here.

(It's also possible to cancel the color charge of a quark using the complementary color charges of two other quarks. This is what happens for baryons, including the proton and neutron—as opposed to mesons, based on quark-antiquark. The principle is the same.)

Third Idea: Einstein's Second Law

So there are two competing effects, which pull in opposite directions. To cancel the field disturbance accurately and minimize that energy cost, Nature wants to localize the antiquark on the quark. But to minimize the quantum-mechanical cost of localizing a position, Nature wants to let the antiquark wander a bit.

Nature compromises. She finds ways of striking a balance between the demands of the gluon fields that don't want to be disturbed, and those of the quarks and antiquarks that want to roam free. (You might think of a family gathering where the gluon fields are the old curmudgeons, the quarks and antiquarks are the rambunctious kids, and Nature is the responsible adult.)

As with any compromise, the result is—well, a compromise. Nature can't make both energies zero simultaneously. So the total energy won't be zero.

Actually there can be different accommodations that are more-or-less stable. Each will have its own nonzero energy E. And thus, according to Einstein's second law, each will have its own mass, $m = E/c^2$.

And that is the origin of mass. (Or at least of 95% of the mass of normal matter.)

Scholium

Such a climax deserves some commentary. Indeed, it deserves a scholium, which is simply Latin for "commentary" but sounds much more impressive.

1. Nothing in this account of the origin of mass referred to, or depended on, the quarks and gluons having any mass. We really do get Mass Without Mass.
2. It wouldn't work without quantum mechanics. You *cannot* understand where your mass comes from if you don't take quantum mechanics into account. In other words, without quantum mechanics you're doomed to be a lightweight.
3. A similar mechanism, though much simpler, works in atoms. Negatively charged electrons feel an attractive electric force from the positively charged nucleus. From that point of view, they'd like to snuggle right on top of it. Electrons are wavicles, though, and that inhibits them. The result, again, is a series of possible compromise solutions. These are what we observe as the energy levels of the atom.
4. The title of Einstein's original paper was a question, and a challenge:

Does the Inertia of a Body Depend on Its Energy Content?

If the body is a human body, whose mass overwhelmingly arises from the protons and neutrons it contains, the answer is now clear and decisive. The inertia of that body, with 95% accuracy, *is* its energy content.

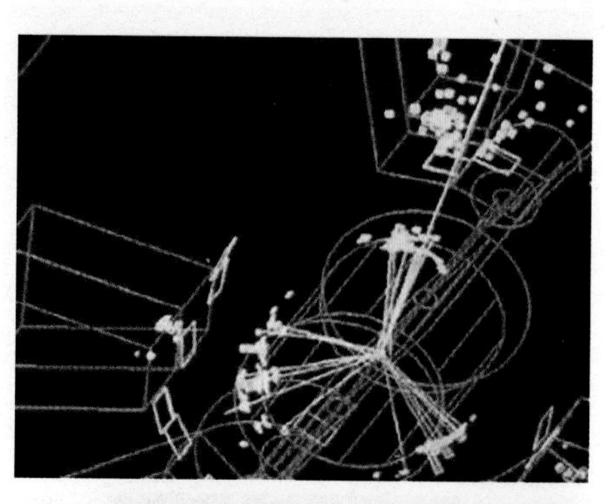

Plate 1 Photograph taken at the Large Electron-Positron collider (LEP) that operated at CERN, near Geneva, through the 1990s. The jets of particles that emerge from this collision follow the flow patterns predicted theoretically for a quark, an antiquark, and a gluon. Jets give operational meaning to those entities, which cannot be observed as particles in the usual sense.

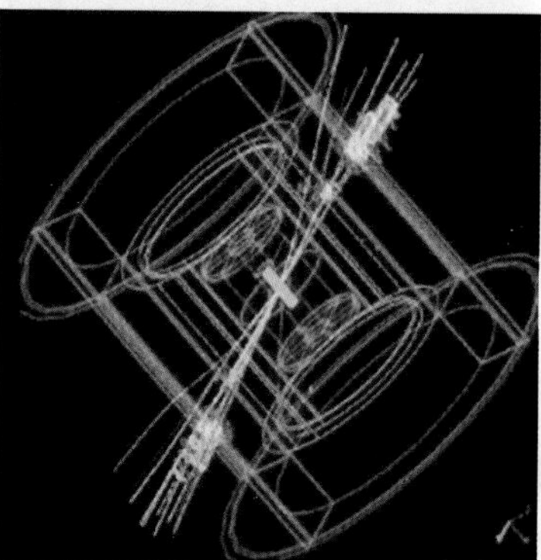

Plate 2 A two-jet process, which we interpret as manifestation of a quark and an antiquark.

a b

Plate 3 Symmetry in science, art, and reality, illustrated through icosahedra. a. An icosahedron has 20 equal sides; all are equilateral triangles. An icosahedron supports 59 distinct symmetry operations; that is, there are 59 distinct rotations that take an icosahedron into itself (while interchanging some of the sides). This compares with 2 distinct symmetry operations for an equilateral triangle. Metaphorically, the symmetry of QCD stands to that of QED as the symmetry of an icosahedron stand to that of a single triangle. b. Enormous symmetry allows one to specify elaborate structures using simple components, a feature viral DNA (or RNA) exploits. Shown here is a virus for the common cold. Note the similarity to a.!

Plate 4 Deep structure of the quantum Grid. This is a typical pattern of activity in the gluon fields of QCD. These patterns of activity are at the heart of our successful computation of hadron masses, as discussed in Chapter 9, so we can be confident that they correspond to reality. This beautiful image was computed by Derek Leinweber, University of Adelaide.

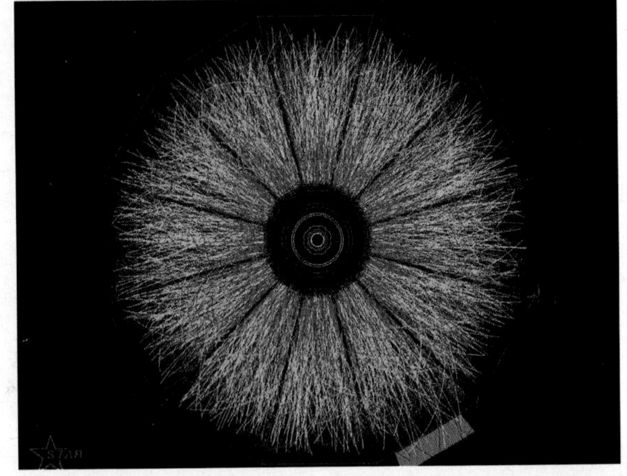

Plate 5 End result of a heavy ion collision—a miniature version of the big bang.

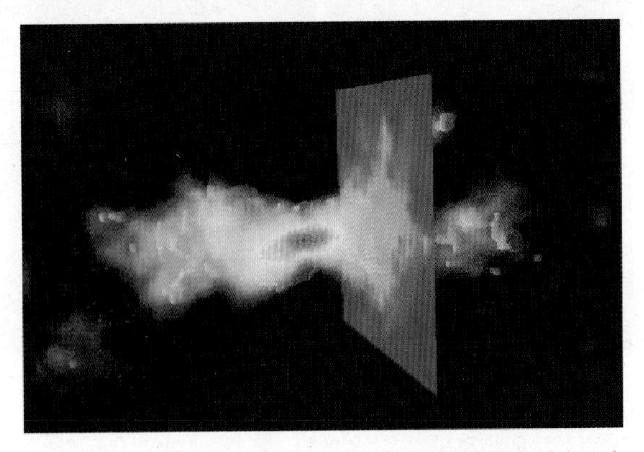

Plate 6 A disturbance in the Grid. At the left, a quark and an antiquark have been injected. They soon establish a dynamic equilibrium, with the energy of the disturbance confined to a small spatial region, moving through time. The Grid fluctuations have been averaged over, leaving only the net distribution of excess energy. By taking a slice, we find the energy distribution inside the particle being reproduced: in this case, a π meson. The total energy gives the mass of the π meson, according to Einstein's second law.

Plate 7 The Siren's alluring song asks us to leave comfortable certainty behind and to meet her on a dubious shore. In return, she promises beauty and illumination. Is she teaching, or teasing?

Plate 8 View of the LHC from the air. The Jura Mountains and Lake Geneva frame a mystic scene. Some image processing has been committed here; in reality the machine is underground.

Plate 9 The ATLAS detector for the LHC, in an early stage of construction. In the final, operational form of the detector, this gigantic framework gets densely packed with magnets, sensors, and ultrafast electronics. This is what it takes to make a camera capable of resolving times of order 10^{-27} second and distances of 10^{-17} centimeter!

Plate 10 Darkness visible. Dark matter does not emit light; it's "seen" only through its gravitational influence on the motion of ordinary matter. Through image processing we can let our eyes see the world as gravitons do. This ROSAT picture shows confined hot gas highlighted in false purple color. It provides clear evidence for gravity exceeding that exerted by the galaxies inside. The extra gravity is attributed to dark matter. Ideas to improve the equations of physics predict new forms of matter whose properties make them good dark matter candidates. Soon we may learn which, if any, of those ideas correspond to reality.

Plate 11 The author with noted blogger Betsy Devine inside a piece of the other main detector for the LHC, whose acronym is CMS—for Compact (!) Muon Solenoid.

11

Music of the Grid: A Poem in Two Equations

The masses of particles sound the frequencies with which space vibrates, when played. This Music of the Grid betters the old mystic mainstay, "Music of the Spheres," both in fantasy and in realism.

LET US COMBINE Einstein's second law

$$m = E/c^2 \tag{1}$$

with another fundamental equation, the Planck-Einstein-Schrödinger formula

$$E = h\nu \tag{2}$$

The Planck-Einstein-Schrödinger formula relates the energy E of a quantum-mechanical state to the frequency ν at which its wave function vibrates. Here h is Planck's constant. Planck introduced it in his revolutionary hypothesis (1899) that launched quantum theory: that atoms emit or absorb light of frequency ν only in packets of energy $E = h\nu$. Einstein went a big step further with his photon hypothesis (1905): that light of frequency ν is always organized into packets with energy $E = h\nu$. Finally Schrödinger made it the basis of his basic equation for wave functions—the

Schrödinger equation (1926). This gave birth to the modern, universal interpretation: the wave function of any state with energy E vibrates at a frequency ν given by $\nu = E/h$.[1]

By combining Einstein with Schrödinger we arrive at a marvelous bit of poetry:

$$(*) \quad \nu = mc^2/h \quad (*)$$

The ancients had a concept called "Music of the Spheres" that inspired many scientists (notably Johannes Kepler) and even more mystics. Because periodic motion (vibration) of musical instruments causes their sustained tones, the idea goes, the periodic motions of the planets, as they fulfill their orbits, must be accompanied by a sort of music. Though picturesque and soundscape-esque, this inspiring anticipation of multimedia never became a very precise or fruitful scientific idea. It was never more than a vague metaphor, so it remains shrouded in quotation marks: "Music of the Spheres."

Our equation (*) is a more fantastic yet more realistic embodiment of the same inspiration. Rather than plucking a string, blowing through a reed, banging on a drumhead, or clanging a gong, we play the instrument that is empty space by plunking down different combinations of quarks, gluons, electrons, photons, . . . (that is, the Bits that represent these Its) and let them settle until they reach equilibrium with the spontaneous activity of Grid. Neither planets nor any material constructions compromise the pure ideality of our instrument. It settles into one of its possible vibratory motions, with different frequencies ν, depending on how we do the plunking, and with what. These vibrations represent particles of different mass m, according to (*). The masses of particles sound the Music of the Grid.

1. Attentive readers will recognize this as Schrödinger's second law.

12

Profound Simplicity

Our best theories of the physical world appear complicated and difficult because they are *profoundly* simple.

EINSTEIN IS OFTEN QUOTED for his advice to "Make everything as simple as possible, but not simpler." After studying either Einstein's general relativity, or his theory of fluctuations in statistical mechanics—two of his more intricate creations—you might well wonder whether he heeded his own advice. Certainly those theories are not "simple" in the usual sense of the word.

Modern physicists consider QCD an almost ideally simple theory, yet we've seen how complicated it is to describe QCD in everyday words, and how challenging it is to work with (and *not* solve) that theory. Like Bohr's profound truth, profound simplicity contains an element of its opposite, profound complexity. This is a paradox, but its resolution is profoundly straightforward, as we'll now explore.

Perfection Supporting Complexity: Salieri, Joseph II, and Mozart

I learned what perfection means from the notoriously mediocre composer Antonio Salieri.[1] In one of my favorite scenes from one

1. Salieri's mediocrity is debated by serious music critics. Regardless, he's *notorious* for being mediocre.

of my favorite movies, *Amadeus*, Salieri looks with wide-eyed astonishment at a manuscript of Mozart's and says, "Displace one note and there would be diminishment. Displace one phrase and the structure would fall."

In this, Salieri captured the essence of perfection. His two sentences define precisely what we mean by perfection in many contexts, including theoretical physics. You might say it's a perfect definition.

A theory begins to be perfect if any change makes it worse. That's Salieri's first sentence, translated from music to physics. And it's right on point. But the real genius comes with Salieri's second sentence. A theory becomes perfectly perfect if it's impossible to change it significantly without ruining it entirely—that is, if changing the theory significantly reduces it to nonsense.

In the same movie, Emperor Joseph II offers Mozart some musical advice: "Your work is ingenious. It's quality work. And there are simply too many notes, that's all. Just cut a few and it will be perfect." The emperor was put off by the surface complexity of Mozart's music. He didn't see that each note served a purpose—to make a promise or fulfill one, to complete a pattern or vary one.

Similarly, at first encounter people are sometimes put off by the superficial complexity of fundamental physics. Too many gluons!

But each of the eight color gluons is there for a purpose. Together, they fulfill complete symmetry among the color charges. Take one gluon away, or change its properties, and the structure would fall. Specifically, if you make such a change, then the theory formerly known as QCD begins to predict gibberish; some particles are produced with negative probabilities, and others with probability greater than 1. Such a perfectly rigid theory, one that doesn't allow consistent modification, is extremely vulnerable. If any of its predictions are wrong, there's nowhere to hide. No fudge factors or tweaks are available. On the other hand, a perfectly rigid theory, once it shows significant success, becomes very powerful indeed. Because if it's approximately right and can't be changed, then it must be exactly right!

Salieri's criteria explain why symmetry is such an appealing principle for theory building. Systems with symmetry are well on the path to Salieri's perfection. The equations governing different objects and different situations must be strictly related, or the symmetry is diminished. With enough violations all pattern is lost, and the symmetry falls. Symmetry helps us make perfect theories.

So the crux of the matter is not the number of notes or the number of particles or equations. It is the perfection of the designs they embody. If removing any one would spoil the design, then the number is exactly what it should be. Mozart's answer to the emperor was superb: "Which few did you have in mind, Majesty?"

Profound Simplicity: Sherlock Holmes, Newton Again, and Young Maxwell

One sure way to avoid perfection is to add unnecessary complications. If there are unnecessary complications, they can be displaced without diminishment and removed without destruction. They also distract, as in this story told of master sleuth Sherlock Holmes and his friend Dr. Watson:

> Sherlock Holmes and Dr. Watson were on a camping trip. After pitching tent beneath a skyful of stars, they went to sleep. In the middle of the night Holmes shook Watson awake, and asked him, "Watson, look up at the stars! What do they tell us?"
>
> "They teach us humility. There must be millions of stars, and if even a small fraction of those have planets like Earth, there will be hundreds of planets with intelligent beings. Some of them are probably wiser than we are. They may be looking through their great telescopes down at Earth as it was many thousands of years ago. They may be wondering whether intelligent life will ever evolve here."
>
> After a moment, Holmes replied, "Actually, Watson, those stars are telling us that someone has stolen our tent."

Passing from the ridiculous to the sublime, you may recall that Sir Isaac Newton was not satisfied with his theory of gravity, which featured forces acting through empty space. But because that theory agreed with all existing observations and he could not discover any concrete improvement, Newton put his philosophical reservations aside and presented it unadorned. In the concluding General Scholium to his *Principia*, he made a classic declaration:[2]

> I have not as yet been able to discover the reason for these properties of gravity from phenomena, and I do not feign hypotheses. For whatever is not deduced from the phenomena must be called a hypothesis; and hypotheses, whether metaphysical or physical, or based on occult qualities or mechanical, have no place in experimental philosophy.

The key phrase "I do not feign hypotheses" is *Hypothesis non fingo* in the original Latin. *Hypothesis non fingo* is the legend that Ernst Mach enshrined below the portrait of Newton in his influential *Science of Mechanics.* It is famous enough to have its own Wikipedia entry. It means, simply, that Newton refrained from loading his theory of gravity with speculation free of observable content. (In his private papers, however, Newton worked obsessively to try to discover evidence for a medium filling space.)

Of course, the easiest way to avoid unnecessary complications is to say nothing at all. To avoid that pitfall, we need a dose of young Maxwell. According to an early biographer, as a small boy he was always asking, "in the Gallowegian accent and idiom," "What's the go o' that?" and, on receiving an unsatisfactory answer, asking, "But what's the *particular* go o' that?"

In other words, we must be ambitious. We must keep addressing new questions and strive for specific, quantitative answers. The phrase *scientific revolution* has been used for so many things that it has been devalued. The emergence of ambition to make precise mathematical world-models, and faith that one could succeed, was the decisive, inexhaustible Scientific Revolution.

---

2. Warning: may induce déjà vu. I quoted this before, in Chapter 8.

There is creative tension between the conflicting demands of economizing on assumptions and providing *particular* answers to many questions. *Profound* simplicity is stingy on the input side, generous on the output side.

Compression, Decompression, and (In)Tractability

Data compression is a central problem in communication and information technology. I think it gives us a fresh and important perspective on the meaning and importance of simplicity in science.

When we transmit information, we want to take best advantage of the available bandwidth. So we boil down the message, removing redundant or inessential information. Acronyms such as MP3 and JPEG are familiar to users of iPods and digital cameras; MP3 is an audio compression format, and JPEG is an image compression format. Of course, the receiver at the other end has to take the boiled-down data and unpack it to reproduce the intended message. When we want to store information, similar problems arise. We want to keep the data compact, but ready to unfold.

From a larger perspective, many of the challenges humans face in making sense of the world are data compression problems. Information about the external world floods our sensory organs. We must fit it into the available bandwidth of our brains. We experience far too much to keep accurate memory of it all; so-called photographic memory is rare and limited, at best. We construct working models and rules of thumb that allow us to use small representations of the world, adequate to function in it. "There's a tiger coming!" compresses gigabytes of optical information, plus perhaps megabytes of audio from the tiger's roar, and maybe even—this means trouble—a few kilobytes of her odor and the wind she stirs up, into a tiny message. (For experts: 23 bytes in ASCII.) A lot of information has been suppressed, but we can unfold some very useful consequences from the little that's there.

Constructing *profoundly* simple theories of physics is an Olympian[3] game in data compression. The goal is to find the shortest possible message—ideally, a single equation—that when unpacked produces a detailed, accurate model of the physical world. Like all Olympian games, this one has rules. Two of the most important are

- Style points are deducted for vagueness.
- Theories that make wrong predictions are disqualified.

Once you understand the nature of this game, some of its strange features become less mysterious. In particular: for the ultimate in data compression, we must expect tricky and hard-to-read codes. For example, consider "Take this sentence in English." Eliminating the vowels, we make it shorter:

Tk ths sntnc n nglsh.

This is harder to read, but there's no real ambiguity about what sentence it represents. According to the rules of the game, it's a step in the right direction. We might go further, eliminating the spaces:

Tkthssntncnnglsh.

That starts to get more questionable. It could be mistaken for

Took those easy not nice nine ogles, he.

Of course, English is so quirky that this kind of code loses heavily on style points, for vagueness. It's hard to be sure exactly what

---

3. It's not in the Olympics, of course, so it's not an Olympic event. But as a challenge worthy of the Greek gods and goddesses, it's Olympian.

counts as a legitimate sentence. In the game of profound simplicity, we must do our decompression using precisely defined mathematical procedures. But as this simple example suggests, we must expect that short codes will be less transparent than the original message, and that decoding them will require cleverness and work.

After centuries of development, the shortest codes could become quite opaque. It could take years of training to learn how to use them—and hard work to read any particular message. And now you understand why modern physics looks the way it does!

Actually, it could be a lot worse. The general problem of finding the optimal way to compress an arbitrary collection of data is known to be unsolvable. The reason is closely related to Gödel's famous Incompleteness Theorem, and (especially) to Turing's demonstration that the problem of deciding whether a program will send a computer into an infinite loop is unsolvable. In fact, looking for the ultimate in data compression runs you straight into Turing's problem: you can't be sure whether your latest wonderful trick for constructing short codes will send the decoder into an infinite loop.

But Nature's data set seems far from arbitrary. We've been able to make short codes that describe large parts of reality fully and accurately. More than this: in the past, as we've made our codes shorter and more abstract, we've discovered that unfolding the new codes gives expanded messages, which turn out to correspond to new aspects of reality.

When Newton encoded Kepler's three laws of planetary motion into his law of universal gravity, explanations of the tides, the precession of the equinoxes, and many other tilts and wobbles tumbled out. In 1846, after almost two centuries of triumph upon triumph for Newton's gravity, small discrepancies were showing up in the orbit of Uranus. Urbain Le Verrier found that he could account for these discrepancies by assuming the existence of a new planet. And lo and behold, when observers turned their telescopes where he suggested they look, there was Neptune! (Today's dark-matter problem is an uncanny echo, as we'll see.)

PART II

The Feebleness of Gravity

IN ASTRONOMY, gravity is the most important force. But fundamentally, acting between elementary particles, gravitational forces are *ridiculously* small compared to electric or strong forces. That disparity poses a big challenge to the ideal of a unified theory, which seeks to put all the forces on the same footing. Our new understanding of the origin of mass suggests an answer.

13

Is Gravity Feeble? Yes, in Practice

When fairly compared, in its action between basic particles, gravity is *ridiculously* feebler than the other fundamental forces.

IF YOU'VE JUST STRUGGLED to haul your body out of bed, or if you've just thankfully collapsed into an easy chair with a good book after a long day, you might find it hard to accept that gravity is feeble. Nevertheless, at a fundamental level, it is. *Ridiculously* feeble.

Here are some comparisons.

Atoms are held together by electric forces. There is electrical attraction between the positively charged atomic nucleus and the negatively charged electrons. Let's imagine that we could turn off the electric forces. There would still be gravitational attraction. How tightly could gravity hold the nucleus and electrons together? How big would a gravitationally bound atom be? As big as a flea? No. A mouse? No. A skyscraper? No, keep going. Earth? Not even close. An atom held together by gravity would have *a hundred times the radius of the visible Universe.*

The deflection of light by the Sun is a celebrated gravitational effect. Its observation, by the British expedition of 1919, was a triumph for general relativity, and established Einstein as a world celebrity. The whole Sun, acting on a nearby photon, deflected its path by 1.75 minutes of arc—about 3% of 1 degree. Now compare that to how the strong force acts on gluons. A few quarks deflect the gluon's straight-line path so severely that within the radius of a proton, the gluon turns around completely and stays inside.

We can also do a numerical comparison. Because both electric and gravitational forces fall off in the same way with distance (namely, as the inverse square), we'll get the same ratio at any distance. Let's compare the electric to the gravitational force between a proton and an electron. The electrical force is about 10,000,000,000,000,000,000,000,000,000,000,000,000,000 times stronger. In scientific notation, that's 10^{40}. (You see why scientists prefer scientific notation.) "Bogus!" carps the critic. "Protons are complicated objects. You should be comparing the forces between basic, elementary objects." Fine, wise guy—that only makes it worse! If we compare the forces between electrons, we get an even bigger number—about 10^{43}—because an electron's mass is smaller than a proton's, but the magnitude of its electric charge is the same.

When you rise out of bed, you overcome the gravitational pull of the entire Earth, using a small part of the chemical energy derived from last night's dinner. As anyone who's tried to burn calories by fighting gravity (lifting weights, doing calisthenics) can attest, gravity doesn't put up much of a fight—a few calories go a very long way.

Another measure of the feebleness of gravity: Electromagnetic radiation is the workhorse of modern astronomy, from radio dishes to optical telescopes to x-ray satellites. It is also the workhorse of modern communications technology, from conventional radio to satellite (microwave dishes) to optical fibers. Gravitational radiation, by contrast, has never yet been detected, despite heroic efforts.

Gravity is the dominant force in astronomy, but only by default. Other interactions are far stronger, but they feature both attractions and repulsions. Normally matter reaches an accurate equilibrium, with the forces cancelled. Temporary imbalances (*small* ones) among electric forces lead to lightning storms; small temporary imbalances among strong forces induce nuclear explosions. Gross breakdowns of equilibrium cannot stand. Gravity, however, is always attractive. Though feeble at the level of individual basic particles, gravitational forces inexorably add up. The meek inherit the cosmos.

14

Is Gravity Feeble? No, in Theory

Gravity is a universal force. It shapes the basic structure of space and time. It is fundamental. So we should use gravity as the measure of other things, and not use other things as the measure of gravity. Gravity, therefore, can't be feeble in an absolute sense—it just is what it is. The fact that gravity *appears* feeble is confounding to theory. It also puts a major hurdle on the path to unification.

EINSTEIN'S THEORY OF GRAVITY, general relativity, ties the existence of gravity to the structure of space and time. The effect we see as the force of gravity, according to this theory, is simply bodies doing their best to travel in straight lines through the curved landscape of space-time. Bodies also cause space-time to curve. The curvature caused by body B affects the motion of body A to produce what in Newtonian language we would call the "force of gravity."[1]

A far-reaching consequence of Einstein's picture of gravity is the *universality* of that force. Any body, doing its best to travel in a straight line through curved space-time, will follow the same path any other body would. The best path is determined by the curvature of space-time, not by any specific property of the body.

1. More precisely, Newton's theory describes the results of general relativity *approximately*. Newton's theory works best when the bodies are slowly moving, compared to the speed of light, and are not too large or dense.

In fact the observed universality of gravity was a big part of what led Einstein to his theory. In Newton's account of gravity, the universality was an unexplained coincidence (or rather an infinite number of coincidences, one for each body). On the one hand, the gravitational force felt by a body is proportional to its mass. On the other, the acceleration a body feels in response to a given force is *inversely* proportional to mass. (That's Newton's second second law of motion. His original second law of motion is $F = ma$; this is $a = F/m$.) Clapping those two hands together, we discover that the gravitational acceleration of a body—the actual disturbance of its motion—doesn't depend on its mass at all!

And that is what's observed: motion independent of mass. The observed behavior is *universal*: all bodies accelerate in the same way under gravity. But in Newton's account there's no reason why it had to occur. It's another of those things that works in practice but not in theory. The gravitational force on a body didn't have to be proportional to the body's mass. We certainly know of forces, such as electric forces, that aren't proportional to mass.

Einstein's theory explains the gravitational "coincidence." Or rather, transcends it: we don't have to speak separately about a force and a response to force, that happen to depend on mass in opposite ways. We just have bodies doing their best to keep going straight through curved space-time. This is profound simplicity at its best.

Universality and Unification

When we come to seek a unified theory including all the forces of nature, the combination of gravity's universality and its (apparent) feebleness poses great difficulties. Here are the alternatives:

- Gravity might be derived from the other fundamental forces. Because it is a small (feeble) effect, maybe gravity is a byproduct, a small residual after the near-cancellation of effects of opposite electric or color charges, or something more exotic.

But then why should it be universal? The other forces are definitely *not* universal: quarks but not electrons feel the strong force; electrons and quarks, but not photons or color gluons, feel the electromagnetic force. It is hard to imagine a simple universal force that has the same consequences for all particles arising from such lopsided ingredients.

- The other forces might be derived from gravity. It's easy to imagine how nonuniversal forces might arise out of a universal one. There could be several *different* solutions of the universal equations with energy concentrated in small regions of space; we'd interpret those solutions as particles with different properties. (Apparently Einstein himself had hopes of constructing a theory of matter along these lines.) But it is hard to see how an extravagantly feeble force can spin off much larger ones.
- All the forces might appear on the same footing, as different aspects—perhaps related by symmetry—of a single whole, like different sides of a die. But again, it is hard to make this idea consistent with gravity being much feebler than the other forces.

Turning it around: Faith in the possibility of unification drives us into a state of denial. We can't accept that gravity really is feeble, even though it appears that way. Appearances—or rather, our interpretation of them—must be deceptive.

15

The Right Question

Theoretically, gravity shouldn't be feeble. In practice, it is. The core of this paradox is that gravity certainly *appears* feeble *to us*. What's wrong with us?

We measure gravity through its effect on matter. The strength of the gravitational force we observe is proportional to the mass of the bodies we use to observe it. The mass of those bodies is dominated by the mass of the protons and neutrons they're made from.

So if gravity *appears* feeble—as we've seen it does—we can either blame gravity itself for being wimpy, or blame the protons (and neutrons) for being lightweights.

High theory suggests that we should regard gravity as fundamental. From that perspective, gravity just is what it is—it can't be explained in terms of anything simpler. So if we're to reconcile theory and practice, the question we must answer is

Why are protons so light?

Asking the right question is often the crucial step toward understanding. Good questions are questions we can come to grips with. Because we've developed a profound understanding of the origin of the proton's mass, "Why are protons so light?" is a question we're ready for.

16

A Beautiful Answer

Why are protons so light? Because we understand how the mass of a proton arises, we can give a beautiful answer to that question. The answer removes a major barrier to a unified theory of the forces, and encourages us to seek such a theory.

LET'S BRIEFLY RECALL how the proton got its mass, with an eye toward finding something in the process that makes the mass small. (This recapitulates part of Chapter 10.)

A proton's mass is a compromise between two conflicting effects. The color charge carried by quarks disturbs the gluon fields around them. The disturbance is small at first but grows as you get farther from the quark. These disturbances in the gluon field cost energy. The stable states will be those with the smallest possible energy, so we have to cancel these costly disturbances. The disturbing influence of the quark's color charge can be nullified by an antiquark of the opposite charge nearby, or—the way it's done in protons—by two additional quarks with complementary colors. Put the nullifying quarks positioned right on top of the original quark, and there'd be no disturbance left. That, of course, would lead to the (null) disturbance with the lowest possible energy (zero).

Quantum mechanics, however, imposes a different energetic cost, which forces a compromise. Quantum mechanics says that a quark (or any other particle) does not have a definite position. It has a spread of possible positions, described by its wave function. We sometimes speak of "wavicles" instead of particles, to emphasize that fundamental aspect of quantum theory. To force a waviquark into a state with a small spread in positions, we must allow it a large energy. In short, it takes energy to localize quarks. The complete nullification we considered in the previous paragraph would require that the nullifying quarks have precisely the same positions as the original quark. That won't fly, because its cost in localization energy is prohibitive.

So there must be a compromise. In the compromise solution, there will be some residual energy from the not-completely-canceled disturbance in the gluon fields, and some residual energy from the not-quite-completely-unlocalized positions of the quarks. From the total E of these energies the proton mass arises, according to Einstein's second law $m = E/c^2$.

In this account, the newest and trickiest element is the way the disturbance in the gluon field grows with distance. It is closely related to asymptotic freedom, a discovery that recently got three lucky people Nobel Prizes. Asymptotic freedom is a subtle feedback effect from virtual particles, as I explained earlier. It can be thought of as a form of "vacuum polarization," in which the entity we call empty space, the Grid, antiscreens an imposed charge. *The Grid Strikes Back*, *The Runaway Grid*, *Grid Gone Wild*—it has the makings of a thinking person's horror movie.

But the reality is subdued. Antiscreening builds up gradually, especially at first. If the seed (color) charge is small, its effect on the Grid starts out small. The Grid itself, by antiscreening, builds up the effective charge, so the next step in the buildup is a little quicker, and so on. Eventually, the disturbance grows large and threatening and must be canceled off. But that might take a while—that is, you might have to get rather far from the seed quark before it happens.

If the disturbance is slow to build, then the pressure to localize the nullifying quarks is correspondingly mild. We don't have to localize very strictly. Thus the energies involved in both the disturbance and the localization are small—and therefore so is the mass of the proton.

And that's why protons are so light!

What I've just given you is what we call a *hand-waving explanation.* You couldn't see me, but while I was typing it I kept interrupting myself to sketch out clouds with my hands, showing my Italian side. Feynman was famous for his hand-waving arguments. Once he explained his theory of superfluid helium to Pauli using such arguments. Pauli, a tough critic, was unconvinced. Feynman kept at it, and Pauli stayed unconvinced, until Feynman, exasperated, asked, "Surely you can't believe that everything I've said is wrong?" To which Pauli replied, "I believe that everything you've said is not even wrong."

To make an explanation that might be wrong we have to get much more specific. When we say protons are light, how light is light? What are the numbers? Can we really explain the *ridiculous* feebleness of gravity, which, you'll remember, involved fantastically small numbers?

Pythagoras' Vision, Planck's Units

Suppose you had a friend in the Andromeda galaxy whom you could contact only by text-messaging. How would you transmit your vital statistics—your height, weight, and age? This friend doesn't have access to Earth's rulers or scales or clocks, so you can't just say, "I'm so-and-so inches tall, such-and-such pounds, and this-and-that years old." You need universal measures.

In 1899 and 1900, Max Planck was deeply immersed in the research that inaugurated the quantum theory. The climax came in December 1900, when he introduced the famous constant *h*—Planck's constant—that appears in the fundamental equations of quantum mechanics we use today. Just before that, he gave an address to the august Prussian Academy of Sciences in Berlin, in which he posed essentially the question above. (Though he didn't phrase it in terms of text-messaging.) He called it the challenge of defining *absolute units.* What excited Planck about his research was not any sense that he might unlock the secrets of the atom, overthrow classical logic, or level the foundations of physics. All that came much later, and from others. What excited Planck was that he saw a way to solve the problem of absolute units.

The problem of absolute units might sound academic, but it is close to the hearts of philosophers, mystics, and philosophically minded scientific mystics.

The manifesto of twentieth- (and twenty-first-) century post-classical physics was issued long before Planck, in around 600 BCE, when Pythagoras of Samos proclaimed an awesome vision. By studying the notes sounded by plucked strings, Pythagoras discovered that the human perception of harmony is connected to numerical ratios. He examined strings made of the same material, having the same thickness, and under the same tension, but of different lengths. Under these conditions, he found that the notes sound harmonious precisely when the ratio of the lengths of string can be expressed in small whole numbers. For example, the length ratio 2:1 sounds a musical octave, 3:2 a musical fifth, and 4:3 a musical fourth. The maxim "All things are number" sums up his vision.

At this remove, it's hard to be sure exactly what Pythagoras had in mind. Probably part of it was a form of atomism, based on the idea that you could build up shapes from numbers. Today's terminology of squares and cubes of numbers descends from that shape building. Our construction of "Its from Bits" richly fulfills the promise that "Some important things are number." In any case, if we take it literally, Pythagoras's maxim surely goes too far. Abstract numbers such as "3" don't have a length, a mass, or a duration in time. Numbers by themselves can't provide physical units for measurement; they can't make rulers or scales or clocks.

Planck's problem of absolute units takes aim at precisely this issue. In this digital age we are used to the idea that information, as it appears in text-messaging, can be encoded in a sequence of numbers (indeed, 1s and 0s). So Planck was asking, in effect: Are numbers sufficient, if not to construct then at least to *describe* every physically meaningful aspect of a material body—in other words, "all things" about it? Specifically, can we convey measures of length, mass, and time using just numbers?

Planck noted that although the Andromedans wouldn't have access to our rulers, scales, or clocks, they would have access to our physical laws, which are the same as theirs. They could measure, in particular, three universal constants:

c: The speed of light.
G: Newton's gravitational constant. In Newton's theory, this is a measure of the strength of gravity. To be precise, in Newton's law of gravity, the gravitational force between bodies of masses m_1, m_2 separated by distance r is $Gm_1 m_2/r^2$.
h: Planck's constant.

(Actually Planck used a slightly different quantity from the modern *h*, which he hadn't invented yet.)

From these three quantities, by taking powers and ratios, one can manufacture units of length, mass, and time. They are called *Planck units.* Here they come:

L_P: The Planck length. Algebraically, it is $\sqrt{\frac{hG}{c^3}}$. Numerically, it is 1.6×10^{-33} centimeter.
M_P: The Planck mass. Algebraically, it is $\sqrt{\frac{hc}{G}}$. Numerically, it is 2.2×10^{-5} gram.
T_P: The Planck time. Algebraically, it is $\sqrt{\frac{hG}{c^5}}$. Numerically, it is 5.4×10^{-44} second.

Obviously Planck units are not very handy for everyday use. The length and times are ridiculously tiny, even for doing subatomic physics. The Planck length, for example, is 1/100,000,000,000,000,000,000 (10^{-20}) times the size of a proton. The Planck mass, 22 micrograms, is not entirely impractical. Vitamin dosages, for example, are often measured in micrograms. So you might go to your health food store and look for pills with a Planck mass of vitamin B12. For fundamental physics, however, the Planck mass is ridiculously big; it is roughly the mass of 10,000,000,000,000,000,000 (10^{19}) protons.

Despite their impracticality, Planck was proud that his units are based on quantities that appear in (presumably) universal physical laws. They are, in his terms, absolute units. You can use them to solve that pressing problem of text-messaging your vital statistics to a friend in Andromeda. You just express your length, mass, and duration in time (that is, your age) as—big!—multiples of the appropriate Planck units.

Over the twentieth century, as physics developed, Planck's construction took on ever greater significance. Physicists came to understand that each of the quantities c, G, and h plays the role of a conversion factor, one you need to express a profound physical concept:

- Special relativity postulates symmetry operations (boosts, a.k.a. Lorentz transformations) that mix space and time. Space and time are measured in different units, however, so for this concept to make sense, there must be a conversion factor between them, and c does the job. Multiplying a time by c, one obtains a length.
- Quantum theory postulates an inverse relation between wavelength and momentum, and a direct proportionality between frequency and energy, as aspects of wave-particle duality; but these pairs of quantities are measured in different units, and h must be brought in as a conversion factor.
- General relativity postulates that energy-momentum density induces space-time curvature, but curvature and energy density are measured in different units, and G must be brought in as a conversion factor.

Within this circle of ideas, c, h, and G attain an exalted status. They are the enablers of profound principles of physics that couldn't make sense without them.

Unification Scorecard

With the help of Planck's units, we can assess how well our understanding of the origin of the proton's mass accounts for the feebleness of gravity, and whether it removes the barrier to unification that gravity's feebleness seemed to present.

If we are going to produce a unified theory in which special relativity, quantum mechanics, and general relativity are primary components, then we should find that the most basic, underlying laws of physics appear natural when expressed in Planck units. No very large or very small numbers should occur in them.

The root of the our trouble with the apparent feebleness of gravity is that the proton mass is very small *in Planck units.* But we've come to understand that the proton mass is not a direct reflection of the most basic laws of physics. It comes from a compromise between gluon field energy and quark localization energy. The *basic* physics behind the proton's mass—the phenomenon that gets the process going—is the underlying basic unit of color charge. The strength of that seed (color) charge determines how fast the growing bloom of gluon field energy becomes threatening; and thus how much of a hit, in quantum localization energy, quarks must take to cancel it; and thus the value of proton mass, according to Einstein's second law.

Is it possible that a reasonable seed charge leads to the actual, very small—in Planck units—value of the proton mass? To answer this question, of course, we have to specify what we consider a *reasonable* value of the seed charge. To measure the strength of the basic seed charge, we need to consider the basic physical effects it causes. We could consider any of several effects: the force it generates, the potential energy, or (for experts) the cross-section. As long as we measure everything at the Planck distance using Planck units, we'll get similar answers whatever measure we use. Since it's the most vivid and familiar effect, let's focus on the force.

So according to Planck, the seed charge is reasonable if it leads to a force between quarks separated by a Planck length that is neither terribly small nor terribly large when measured in Planck units. Of course he would say that. The point is not the prestige of Planck's authority but the ideal his units embody: the ideal that special relativity, quantum mechanics, and gravity (general relativity) can be unified with the other interactions. We're turning it around and asking if, by *assuming* that ideal, we are led to a consistent understanding of why protons are light, and thus also of why gravity is feeble in practice.

Finally, then, it all boils down to a very concrete numerical question: Is the magnitude of the seed strong force between quarks, at the Planck length, expressed in Planck units, close to 1?

To answer that question we must extrapolate the laws of physics we know down to distances far smaller than where they have been checked experimentally. The Planck length is very small. Many things could go wrong. Nevertheless, in the spirit of our Jesuit Credo, "It is more blessed to ask forgiveness than permission," let's just do it.

The required calculation is actually quite a simple one by the standards of modern theoretical physics. We've discussed all the necessary ideas in words. It breaks my heart not to display the algebra, but I'm a merciful man, and besides my publisher warned me against it. So I'll just state the result:

We find that the seed strong force between quarks, at the Planck scale, measured in Planck units, is about 1/25. That's quite an improvement over the 1/10,000,000,000,000,000,000,000,000,000,000,000,000,000 discrepancy we thought we had!

Thus we've explained the (apparent) feebleness of gravity starting from fundamental, new, yet firmly based physics. And we've overcome a major obstacle blocking the path toward a unified theory of the forces.

Next Steps

I hope you'll agree that it's a pretty story, and it hangs together. Declarations of "mission accomplished" have been based on much less.

PART III

Is Beauty Truth?

WE'VE FOUND AN EXPLANATION for the feebleness of gravity that is logical and beautiful. But is it true? To establish its truth and fertility—or not—we need to embed it in a wider circle of ideas, and draw out testable consequences.

Nature seems to be hinting that a unified theory of the fundamental forces is possible. Our explanation of the feebleness of gravity fits very comfortably in that framework. But to consummate the unification fully, in detail, we must postulate the existence of a new world of particles. Some of those particles should materialize at the great LHC (Large Hadron Collider) accelerator, located near Geneva. One of them might also pervade the Universe, providing its dark matter.

17

Unification: The Siren's Song

The known particles and interactions present a fragmentary pattern. An expanded theory, based on the same principles but featuring larger symmetry, brings them together.

We've seen that by following some well-established laws of physics where they lead us, we arrive at a profound explanation of one of the subject's classic problems: Why is gravity so feeble?

Unfortunately, to get to that answer we had to project those well-established laws down to distances far smaller than those we can hope to examine directly. Equivalently,[1] we had to project the laws we have up to energies far larger than any we can hope to examine directly—energies literally "millions of billions" (10^{15}) times those that the latest and greatest, multibillion-euro accelerator LHC is built to achieve. Thus our explanation is firmly based—on an untested foundation!

We need not accept this situation passively. We can search for other ways to access the physics of unification, and to look into ultrashort distances and ultrahigh energies. The direct path is blocked. We cannot, as a practical matter, accelerate particles and smash them together at the required energies. We can, however,

1. We've discussed the close connection between ultrashort distances and ultralarge energies previously. See the endnotes for pointers and some additional comments.

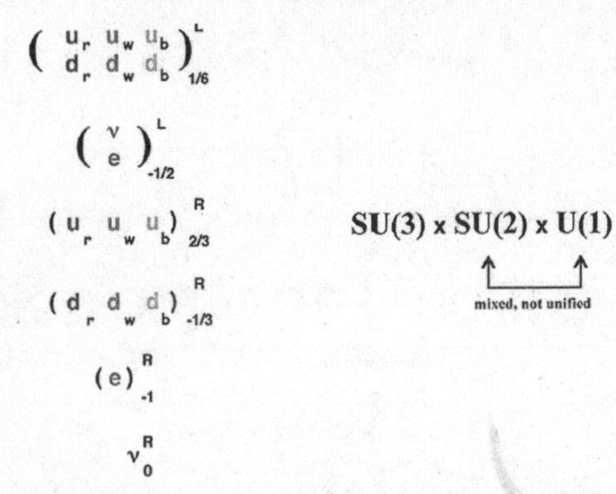

Figure 17.1 The organization of particles and interactions in the Core. What leaps to the eye is that the quarks and leptons fall into six distinct groups, and the interactions fall into three distinct parts.

look for additional signs of unification—unexplained patterns in the world we do have access to.

Such patterns are there. Please take a look at Figures 17.1 and 17.2.

Figure 17.1 presents the organization of particles as we find them—the so-called standard model (including QCD). *Standard model* is a grotesquely modest name for one of humankind's greatest achievements. The standard model summarizes, in remarkably compact form, almost everything we know about the fundamental laws of physics.[2] All the phenomena of nuclear physics, chemistry, materials science, and electronic engineering—it's all there. And unlike Feynman's jocular $U = 0$ or the verbal gymnastics of classical philosophy, this figure comes with definite algorithms for unfolding symbols into a model of the physical world. It allows you to make surprising predictions and to design, for example, exotic lasers, nuclear reactors, or ultrafast ultrasmall computer memories with confidence. Not being grotesquely modest, I will henceforth refer to the standard model as the Core theory.

2. I'll get to the exceptions shortly.

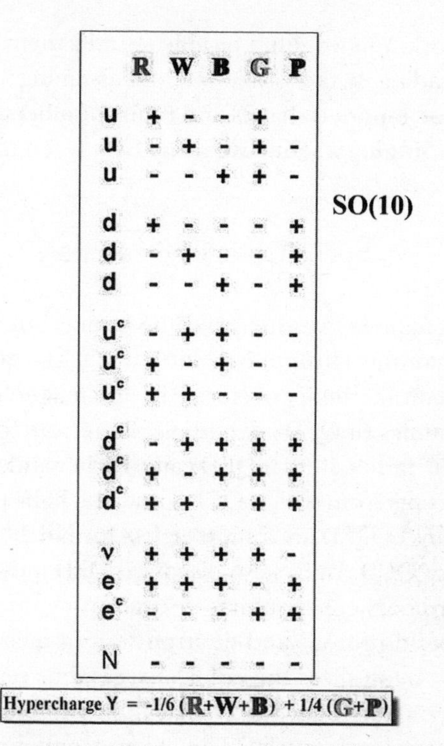

	R	W	B	G	P
u	+	-	-	+	-
u	-	+	-	+	-
u	-	-	+	+	-
d	+	-	-	-	+
d	-	+	-	-	+
d	-	-	+	-	+
u^c	-	+	+	-	-
u^c	+	-	+	-	-
u^c	+	+	-	-	-
d^c	-	+	+	+	+
d^c	+	-	+	+	+
d^c	+	+	-	+	+
ν	+	+	+	+	-
e	+	+	+	-	+
e^c	-	-	-	+	+
N	-	-	-	-	-

Figure 17.2 The organization of the same particles and interactions, plus more, into a unified theory. What leaps to the eye is that the quarks and leptons are unified into one whole, as are the interactions.

It would be hard to exaggerate the scope, power, precision, and proven accuracy of the Core. So I won't even try. The Core is close to Nature's last word. It will provide the core of our fundamental description of the physical world for a long time—possibly forever.

Figure 17.2 presents an organization of the same particles and their properties in a unified theory. It is far from automatic that the (established) Core can be subsumed in the (hypothetical) unified theory. If the lopsided shapes that appear in the Core pattern or the funny numbers that hang from them were different, it wouldn't work. You wouldn't be able to unify them (at least not so neatly). Reading it the other way: by assuming unification, we explain those lopsided shapes and funny numbers.

Nature is singing a seductive song. Let's listen more closely . . .

The Core: Choice Bits

In earlier chapters, we discussed the strong interaction and its theory—quantum chromodynamics, or QCD—quite a lot. The modern quantum theory of electricity and magnetism—quantum electrodynamics, or QED—is both the father and the baby brother of QCD. The father, in that QED came earlier and provided many of the concepts from which QCD grew; the baby brother, in that the equations of QED are a simpler, less formidable version of the equations of QCD. We've also discussed QED quite a lot.

In the ordinary course of nature, the strong interaction's main role is to build protons and neutrons out of quarks and gluons. This almost neutralizes the color charges, but remaining imbalances generate residual forces that bind protons and neutrons together into atomic nuclei. The electromagnetic interaction in turn binds electrons to those nuclei, producing atoms. This almost neutralizes the electric charges, but remaining imbalances generate the residual forces that bind atoms into molecules and molecules into materials. QED also describes light and all its cousin forms of electromagnetic radiation: radio, microwave, infrared, ultraviolet, x-ray, γ-ray.

The third major player in the Core is the weak interaction. Its role in nature is more subtle, but also crucial. The weak interaction performs alchemy. More precisely, it morphs different flavors of quarks into one another, and also different kinds of leptons into one another. In Figure 17.1, the weak interaction makes transformations in the vertical direction. (The strong interaction makes transformations in the horizontal direction.) When you change one of the u quarks in a proton into a d quark, the proton becomes a neutron. So changes wrought by the weak interaction transform an atomic nucleus of one element into the atomic nucleus of another. Reactions based on weak interaction "alchemy" (a more respectable name is nuclear chemistry) can release enormously larger energies than ordinary chemical reactions. Stars live on energy derived from systematically baking protons into neutrons.

Before going into more details about the core of the Core—the strong, electromagnetic, and weak interactions—let me make a few comments about what I'm (temporarily!) leaving out. There are two big items: gravity and neutrino masses.

- As we've already discussed, the apparent feebleness of gravity probably has more to do with our special perspective than with gravity itself. And as we'll see over the next few chapters, Nature encourages us to include gravity with the other interactions, as an equal partner in unification.

  There is no difficulty, in practice, including gravitational interactions into the Core. There is a unique, straightforward way to do it, and it works. (For experts: Use the Einstein-Hilbert action for the metric field, use minimal coupling to the matter fields, and quantize around flat space.) Astrophysicists use general relativity together with the rest of the Core routinely, and quite successfully, in their everyday work. So does anyone who uses GPS.

  In short: The usual convention of separating gravity from the Core is handy, but probably superficial.
- That neutrinos have nonzero mass was established in 1998, although there were hints going back to the 1960s. The values of neutrino masses are very small. The heaviest of the three types of neutrinos has no more than a millionth the mass of the next-lightest particle we know about, the electron. Neutrinos are famously elusive and ghostly. Roughly 50 trillion of them pass through each human body every second, without our noticing. John Updike wrote a poem about neutrinos, which begins:

> Neutrinos they are very small.
> They have no charge and have no mass
> And do not interact at all.
> The earth is just a silly ball
> To them, through which they simply pass

Nevertheless, by heroic efforts experimentalists have been able to study the properties of neutrinos in considerable detail.[3]

The Core was happy with zero neutrino masses, which fit into its structure very naturally. To accommodate nonzero neutrino masses we must add in new particles, with exotic properties, for which there's no other motivation or evidence. When we extend the Core, to make a unified theory, things will look very different. Then we'll recognize the new particles as kin of those we know—prodigals returning home, completing the family. And their exotic behavior will hint of adventures in far-off, romantic locales.

There are also two complications that I'm mostly going to gloss over. They're diversions from my central message, but it would be improper not to mention them. Please don't be intimidated or put off by these superficial complications: we must acknowledge their existence, but we won't allow them to confuse our view.

The first complication is the masses and mixings of gauge bosons. In the basic equations, there are three groups of gauge fields. There are eight color gluon fields, which you've become familiar with. Another three are associated with the weak interaction symmetry. They are called W^+, W^-, and W^0, and they are all symmetric with one another. Finally, there is one isolated "hypercharge" gauge boson B^0. Grid superconductivity gives nonzero masses to the

---

3. Whole books have been written about neutrinos and their properties. (They do interact, after all—just very rarely.) Because the subject is highly technical and somewhat tangential to our main topics, I've been very selective and telegraphic in this discussion. For a few more details, and further references, see the endnotes.

particles created by W^+, by W^-, and by a certain *mixture* of W^0 and B^0. Disturbances in that mixture produce the massive particles called Z bosons. Disturbances in another combination of W^0 and B^0 (for experts: the orthogonal combination) remain massless. That massless combination of W^0 and B^0 is the photon.

To summarize: From the point of view of the mathematics of symmetry, the W^0 and B^0 fields are the most natural. But the disturbances with definite mass, once Grid superconductivity is taken into account, involve mixtures of W^0 and B^0. One type of disturbance is the Z boson, with nonzero mass; the other is the photon, with zero mass.

It is sometimes said the Core unifies electromagnetism and the weak interaction. That is misleading. There are still two distinct interactions involved, associated with different symmetries. They are mixed, rather than unified, in the Core theory.

The second complication is the masses and mixings of quarks and leptons. These particles come in three different varieties, or "families." Thus besides the lightest family, which contains u and d quarks, the electron e, and the electron neutrino ν_e, there are two heavier ones. The second family contains the charm and strange quarks c and s, the muon μ, and the muon neutrino ν_μ. Finally, the third family contains the top and bottom quarks t and b, the tau lepton τ, and the τ neutrino ν_τ.

Like the gauge bosons, all these particles would be massless but for Grid superconductivity. But Grid superconductivity gives them mass[4] and also allows the heavier ones to mix with, and thereby decay into, lighter ones in complicated ways. These masses and mixings are extremely interesting to experts, and understanding their values is an unmet challenge for theoretical physics. Also not understood at all is the simpler question: Why are there three families to begin with?

---

4. The neutrinos are a special case, as we just discussed.

Because I don't have any very good ideas about those issues, I won't waste words by insisting on their details. They would only add distracting noise around the good ideas I do want to discuss. So I'll keep things as simple as possible—or maybe even a little simpler. Tolstoy's *Anna Karenina* famously opens, "All happy families are happy in the same way." That being the case, we'll focus on just one.

Phew! It's a complicated business, getting to simplicity. But after we've put those two odd gifts of gravity and neutrino masses into temporary storage in the attic, tidied up after the mix-ups made by Grid superconductivity, and decided that one family is enough, a clear and uncluttered image emerges. It is what you see in Figure 17.1. This is the core of the Core.

There are three symmetries, $SU(3)$, $SU(2)$, and $U(1)$. They correspond to the strong, weak, and electromagnetic interactions,[5] respectively.

$SU(3)$ is a symmetry among three kinds of color charge, as we've already discussed. It comes together with eight gauge bosons that change or respond to the color charges. It acts in the horizontal direction in Figure 17.1.

$SU(2)$ is a symmetry among two additional kinds of color charges. It acts in the vertical direction in Figure 17.1.

You'll notice that each of the particles on the left is listed twice. Each occurs once in a group with the subscript L and once in a group with the subscript R. These subscripts refer to the *handedness,* or *chirality,* of the particles: L for left-handed, R for right-handed. The handedness of a particle is defined as in Figure 17.3. Left-handed and right-handed particles interact differently. This fact is called parity violation. It was first realized by T. D. Lee and C. N. Yang in 1956, and that discovery earned them a Nobel Prize in the least possible time, in 1957.

---

5. Strictly speaking, electromagnetism is a mixture involving pieces from both $SU(2)$ and $U(1)$, as we just discussed. So the $U(1)$ is not quite electromagnetism. It has its own proper name, hypercharge. But I'll generally use the more familiar, not-quite-pedantically-correct name for it.

Figure 17.3 The handedness, or chirality, of a particle is determined by direction of its spin relative to its direction of motion.

U(1) deals with only one kind of charge. We specify its action on the different particles according to how strongly, and with what sign, its one boson—essentially, the photon—couples to each. The little numbers dangling from each grouping of particles specify exactly that, for the particles in the grouping. For example, the right-handed electron has a -1 because its electric charge is -1 (in units where the proton's charge is $+1$). The biggest grouping, with six members, has *u* and *d* quarks with each of the three color charges. The *u* quarks have electric charge $\frac{2}{3}$, and the *d* quarks have electric charge $-\frac{1}{3}$, so the average electric charge within the group is $\frac{1}{6}$, which is the number you see.

And that's it. As I said before, it would be difficult to overstate the power and scope of the Core. The rules might appear a little complicated at first, but those complications are nothing compared to (for instance) the conjugations of a few irregular verbs in Latin or French. And unlike the latter, the complications of the Core are not gratuitous. They are forced on us by experimental realities.

Critique

The score of Nature's song, as we hear it, is Figure 17.1. We've recorded it, and we've been able to compress the recording into an extremely compact form. It's a great achievement summing up centuries of brilliant work.

Judged by the highest esthetic standards, however, there's plenty of room for improvement. Salieri, looking at this score, would definitely not be moved to exclaim, "Displace one note and there would be diminishment." More likely he'd say, "Interesting sketch, but it needs work."

Or perhaps, knowing that he was viewing the work of a master, Salieri might exclaim, "Nature certainly left Her work with a quirky copyist!"

First of all, there are three unconnected interactions. They are based on the same principles of symmetry and response to charges, but the charges involved fall into three distinct groups that can't be transformed into one another. There are transformations (involving the color gluons of QCD) that transform the red, white, and blue color charges among themselves, and there are separate transformations (involving the *W* and *Z* bosons) that transform the green and purple color charges into one another. And electric charge is a separate thing altogether.

Worse, the different quarks and leptons fall into six unrelated clusters. And the clusters are mostly unimpressive: one contains 6 members, but the others are mere hints of motifs, with 3, 3, 2, 1, and 1 members, respectively. Most discordant are those funny numbers, the average electric charges attached to each cluster. They seem pretty random.

The Charge Account

Fortunately, the Core contains the seeds of its own transcendence. Its ruling principle is symmetry, and symmetry is a concept we can build on by pure thought, just using our noodles. We can play with the equations.

For example, we can imagine that there are transformations that turn strong color charges into weak ones, and vice versa. This will generate bigger clusters of related particles, and maybe they'll click into attractive patterns. In the best case, we might hope that the three distinct symmetry transformations of $SU(3) \times SU(2) \times U(1)$ are different facets of one larger, master symmetry that includes them all.

The mathematics of symmetry is well developed, so there are powerful tools for this kind of pattern recognition task. There aren't many possibilities, so we can try them out systematically.

The master symmetry I find most convincing is based on a group of transformations known as $SO(10)$. All the attractive possibilities are minor variants of this one.

Mathematically, $SO(10)$ consists of rotations in a ten-dimensional space. I should emphasize that this "space" is purely mathematical. It's not a space that you could move around in, even if you were very small. Rather, the ten-dimensional space of $SO(10)$, the master symmetry that absorbs the $SU(3) \times SU(2) \times U(1)$ of the Core—that unifies, in other words, the strong, weak, and electromagnetic interactions—is a home for concepts. In this space, each of the color charges of the Core (red, white, blue, green, and purple) is represented by a separate two-dimensional plane (so there are $5 \times 2 = 10$ dimensions altogether). Because there are rotations that move any plane into any other, the Core charges and symmetries get unified and expanded in $SO(10)$.

Mathematical sophisticates will not find it surprising that symmetries can be combined into larger symmetries. As I said, the tools for doing that are well developed. What is much less automatic, and therefore much more impressive, is that the scattered clusters of quarks and leptons fit together. This is what is displayed in Figure 17.2. It is what I like to call the Charge Account.

In the Charge Account, all the quarks and leptons appear on an equal footing. Any of them can be transformed into any other. They fall into a very specific pattern, the so-called spinor representation of $SO(10)$. When we make separate rotations in the two-dimensional planes, corresponding to the red, white, blue, green, and purple charges, we find in each case that half the particles have a positive unit of charge, half a negative unit. These appear as the + and – entries in the Charge Account. Each possibility for combinations of + and – occurs exactly once, subject to the restriction that the total number of + charges is even.

The electric charges, which within the Core appear to be random decorations, become essential elements in the harmony of unification. They are no longer independent of the other charges. The formula

$$Y = -\frac{1}{3}(R + W + B) + \frac{1}{2}(G + P)$$

expresses electric charge—more precisely, hypercharge—in terms of the others. Thus the transformations associated with electric charge rotation turn each of the first three planes through some common angle, and turn the last two through $\frac{3}{2}$ as big an angle, in the opposite sense.

To accomplish this degree of unification, we must realize that right-handed particles can be regarded as the antiparticles of their own (left-handed) antiparticles. For example, the right-handed electron is the antiparticle of the left-handed positron. These descriptions have the same physical content, because both a particle and its antiparticle are excitations in the same field, and it is the field that appears in the primary equations. Symmetry transformations among the fields relate excitations of the same handedness, so to find all the possible symmetries, we deal with left-handed excitations only (even if that means working with antiparticles).

When we descend from Charge Account to Core, we must realize that complementary color charges cancel. Equal amounts of red, white, and blue charge—or equal amounts of green and purple charge—cancel to nothing. Thus, for instance, the three equal (+) red, white, and blue color charges of the left-handed electron *e* cancel. They cancel in the right-handed electron, which is represented in the Charge Account by its left-handed antiparticle e^c

as well. Electrons of either sort are invisible to the color gluons of QCD. In other words, electrons do not participate in the strong interaction.

The most singular entry in the Charge Account is the last one, *N*. Both its strong color charges and its weak color charges cancel. Thus it is invisible to both the strong and the weak interactions. Its electric charge is also zero. So this particle doesn't respond to any of the conventional Core forces. That makes it hopelessly difficult to detect—worse than neutrinos, which at least participate in the weak interaction. (*N* does both feel and exert gravity, but for practical purposes the gravity of individual particles is ridiculously feeble, as we've discussed.)

Sure enough, *N* has not been observed. How could it be? If we observe it, it can't be *N*, which by definition is unobservable! That "triumph" of theory is hollow, of course. But *N* is welcome for a more positive reason. It is the extra particle ν^R that, when added to the Core, allows the neutrino to have its tiny mass.[6]

The left-hand column of the Charge Account specifies the names of the particles—the quarks and leptons—that we use to construct the Core, and the world. But really, we could delete that column. If we didn't know the names of those particles or anything about their properties, and had only the unlabeled Charge Account, nothing would be lost. We could reconstruct the properties of all the particles, from the information in the Charge Account (and of course their names are just a convenience).

Conversely, if the Core clusters had slightly different shapes, or if the funny numbers that dangled from them were different, the pattern would fail.

The Charge Account maps the mathematically ideal to the physically real. It is fully worthy of Salieri's highest accolade: "Displace one note and there would be diminishment. Displace one phrase and the structure would fall."

---

6. More on this later, in Chapter 21.

The Siren's Song

The Sirens of myth sing entrancing songs from rocky coasts, luring sailors to shipwreck and destruction. Their song promises knowledge of secrets, of the past and of the future. They promise "all that comes to pass on the fertile earth, we know it all!" Jane Ellen Harrison comments, "It is strange and beautiful that Homer should make the Sirens appeal to the spirit, not to the flesh."

We've heard a Siren's song of unification.

18

Unification: Through a Glass, Darkly

The higher symmetry that unifies the basic particles also predicts equality among the different basic interactions. That prediction, on the face of it, is quite wrong. But when we correct for the distorting effect of Grid fluctuations it comes close.

WE'VE HEARD A SIREN'S SONG OF UNIFICATION. Now it's time to open our eyes, to see whether we can navigate the rocky coasts where she dwells.

Symmetry Not

The enhanced symmetry of unification does some great things. It assembles the scattered pieces of the Core into well-proportioned wholes. Once our vision accommodates to that dazzling first impression, however, and we begin to look more carefully, things don't seem right.

In fact, something very basic appears to be wrong. If the strong, weak, and electromagnetic forces are aspects of a common underlying master force, then symmetry requires that they should all have the same strength. But they don't. Figure 18.1 shows this.

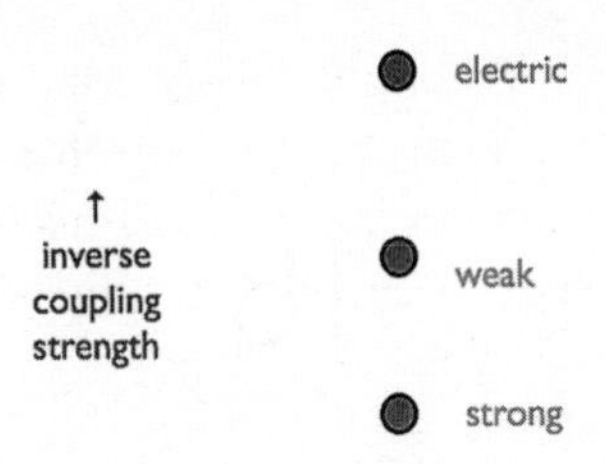

Figure 18.1 Perfect symmetry would require the strong, weak, and electromagnetic forces to have equal strengths. They don't. For later convenience, here I've used the inverse square of the couplings as the quantitative measure of their relative power. So the strong interaction, which is the strongest, appears on the bottom.

There's a reason why the strong interaction is called strong and the electromagnetic interaction isn't. The strong interaction really is stronger! One manifestation of the difference is that atomic nuclei, bound together by strong-interaction forces, are much smaller than atoms, held together by electromagnetic forces. The strong forces hold nuclei together more tightly.

The mathematics of the Core theory enables us to give a precise numerical measure of the relative strength of different interactions. Each of its interactions—strong, weak, electromagnetic—has what we call a coupling parameter, or simply a coupling.

In terms of Feynman graphs, the coupling is a factor by which we multiply each hub. (These universal, overall coupling factors come on top of the purely numerical values of the color or electromagnetic charges of the particular particles involved, as encoded in the Charge Account.) So each time a color gluon appears in a hub, we multiply the contribution of the process depicted by the strong coupling; each time a photon appears, we multiply by the electromagnetic coupling. The basic electromagnetic force comes from exchanging a photon (Figure 7.4), so it has the square of the electromagnetic coupling. Similarly, the basic strong force comes from exchanging a gluon, so it has the square of the strong coupling.

Complete symmetry among the forces requires every hub to be related to every other. It leaves no room for differences among the couplings. The observed differences, therefore, pose a critical challenge to the whole idea of achieving unification through symmetry.

Correcting Our Vision

A great lesson from the Core is that the entity we perceive as empty space is in reality a dynamic medium full of structure and activity. The Grid, as we've called it, affects the properties of everything within it—that is, everything. We see things not as they are, but as through a glass, darkly. In particular, the Grid is aboil with virtual particles, and these can screen or antiscreen a source. That phenomenon, for the strong force, was central to the stories that unfolded in Parts I and II. It occurs for the other forces too.

So the coupling values we see depend on how we look. If we look crudely, we will not discern the basic sources themselves, but will see their image as distorted by the Grid. We will, in other words, see the basic sources mixed together with the cloud of virtual particles that surround them, unresolved. To judge whether perfect symmetry and unity of the forces occurs, we should correct for the distortions.

To see down to the basics, we may need to resolve very short distances and very short times. That's an oft-repeated lesson, from van Leeuwenhoek and his microscopes, to Friedman, Kendall, and Taylor using their ultrastroboscopic nanomicroscope at SLAC to look inside protons, to experimenters using the creative destruction machine LEP to dig into Grid. As we saw in connection with those two recent projects, to resolve extremely short distances and times, where quantum theory comes into play, it's necessary to use probes that actively transfer large amounts of momentum and energy to the object being probed. That's why high-energy accelerators, despite their expense and complexity, are the instruments of choice.

A Near Miss

As we discussed in Chapter 16, the virtual particle clouds can be slow to build. For the cloud around a quark to grow from a reasonable-sized seed to threatening proportions, it had to evolve from the Planck length to the proton's size: a factor of 10^{18} in distance!

Given that experience, we should not be surprised to find that to get to the basics—to see down to the distances where unification might take place—we might need to transfer outlandish amounts of momentum and energy. The next great accelerator, the LHC, will give us ten times better resolution—that is, a factor

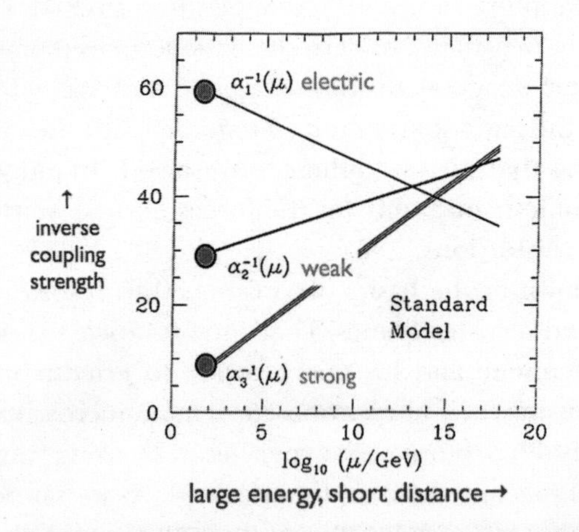

Figure 18.2 Correcting for Grid distortions, to see whether the forces unify. If we plot things in this way—with inverse couplings squared ascending in the vertical direction versus logarithm of energy or (equivalently) of inverse distance in the horizontal direction—then the corrected couplings, viewed at better and better resolution, track straight lines. The size of the experimental errors is indicated by the width of the lines. It almost works, but not quite.

of 10^1—at a cost of roughly *ten billion* euros. And after that, it gets really difficult.

So we have to use our noodles instead. Though not as foolproof, they're relatively cheap, and ready at hand (so to speak). With a few strokes of a pen, we can calculate the effects of Grid distortion and correct for them.

The result is displayed in Figure 18.2.

As Homer Simpson might say:

D'Oh!

It doesn't quite work. Close, but no cigar.

What to do?

19

Truthification

When an attractive idea is close to being right, we try to find ways to make it right. We look for ways to *truthify*.

THE FAMOUS PHILOSOPHER KARL POPPER emphasized the importance of falsifiability in science. The mark of scientific theories, according to Popper, is that they make statements—predictions—that might be false. Is Popper's claim true? Well, can you falsify it?

Maybe it's a profound truth. Reppopism—the opposite of Popperism—says that the mark of a good scientific theory is that you can truthify it. A truthifiable theory might make mistakes, but if it's a good theory they're mistakes you can build on.

In a crucial way, falsifiability and truthifiability are two sides of the same coin. Both value definiteness. The worst kind of theory, on both accounts, is not a theory that makes mistakes. Mistakes, you can learn from. The worst kind of theory is a theory that doesn't even try to make mistakes—a theory that's equally ready for anything. If everything is equally possible, then nothing is especially interesting.

In terms of our Jesuit Credo, "It is more blessed to ask forgiveness than permission," a falsifiable theory asks permission, a truthifiable theory asks forgiveness—and an unscientific theory has no sense of sin.

The ideas of pattern recognition and description compression that we discussed earlier give a different perspective on these issues (and, I think, go deeper). If the processing of every pixel results in a middling shade of gray, no image will emerge from a raw exposure. Similarly, in order to recognize patterns in our exposure to the physical world, against the background of everything that could be imagined, our candidate theories must distinguish possible from impossible (according to the theory). Only then can we color them differently, and only then will our observations give us a picture we can work with—a picture with contrast.

If we are definite and manage to get a lot of pixels right, a useful image can emerge even if there are a few mistakes too. (We can touch it up with Photoshop.) Thus there's a premium on ambition—that is, on bringing a lot of pixels into the picture (or, in our metaphor, a lot of facts to bear)—as well as on accuracy.

Enough metaphors and generalities! On to a case study in truthification.

Upping the Ante: More Unification

Our ambitious attempt to unify the strong, electromagnetic, and weak interactions didn't quite work. We succeeded in producing a theory that was not just falsifiable but outright false. Very scientific, says Sir Karl Popper. But somehow, we are not left feeling gratified.

When such an attractive and nearly successful idea doesn't seem quite right, it makes sense to try to rescue it. We look for ways to truthify it.

Maybe, in our quest for unification, we haven't been ambitious enough. The soul of our unification of different charges is this:

$$\text{electron} \longleftrightarrow \text{quark}$$

$$\text{photon} \longleftrightarrow \text{gluon}$$

20

Unification ♥ SUSY

When we expand the equations of physics to include supersymmetry, we enrich the Grid. Thus we must recalibrate our calculation of how the Grid distorts our view of unification. When we make the correction, a sharp image of unification comes into focus.

BY PERFECTING OUR EQUATIONS, we enlarge the world.

In the 1860s, James Clerk Maxwell assembled the equations for electricity and magnetism as they were understood at the time, and found that they led to contradictions.[1] He saw that he could make them consistent if he added a new term. The new term, of course, corresponds to a new physical effect. Some years before, Michael Faraday (in England) and Joseph Henry (in the United States) had discovered that when magnetic fields change in time, they create electric fields. Maxwell's new term embodied the converse effect: changing electric fields produce magnetic fields. Combining those effects, we get a dramatic new possibility: changing electric fields produce changing magnetic fields, which produce changing electric fields, which produce changing magnetic fields, ... You can have a self-renewing disturbance that takes on a life of its own. Maxwell saw that his equations had solutions of this kind. He could calculate the speed at which these disturbances

1. I mentioned this before, in Chapter 8.

would move through space. And he found that they move at the speed of light.

Being a very bright fellow, Maxwell leaped to the conclusion that these electromagnetic disturbances *are* light. That idea has held up to this day, with many many fruitful applications. It remains the foundation of our deepest understanding of the nature of light. But there's more. Maxwell's equations also have solutions with smaller wavelengths than those of visible light, and with larger wavelengths. So the equations predicted the existence of new kinds of things—new kinds of matter, if you like—that weren't known at the time. These are what we now know as radio waves, microwaves, infrared, ultraviolet, x-rays, γ-rays—each a significant contributor to modern life, each an immigrant from concept world to physical world (c-world to p-world).

In the late 1920s, Paul Dirac was working to improve the equation that describes electrons in quantum mechanics. A few years before, Erwin Schrödinger had formulated an electron equation that worked very well in many applications. But theoretical physicists were not entirely satisfied with Schrödinger's electron equation, because it does not obey the special theory of relativity. It is a quantum-mechanical version of Newton's force law; it obeys the old mechanical relativity rather than Einstein's electromagnetic relativity. Dirac found that to make an equation consistent with special relativity, he had to use a larger equation than Schrödinger's. Like Maxwell's perfected equations for electricity and magnetism, Dirac's perfected equation for electrons had new kinds of solutions: besides the solutions that correspond to electrons moving with different velocities and spinning in different directions, there were others. After some struggles and false starts, and with some help from Hermann Weyl, by 1931 Dirac had deciphered the meaning of these strange new solutions. They represent a new kind of particle, with the same mass as the electron but opposite charge. Just such a particle was discovered soon thereafter, in 1932, by Carl Anderson. We call it the antielectron, or positron. Nowadays we use positrons to monitor what's happening inside brains (via the PET scan, a.k.a. positron-electron tomography).

There are many other recent examples where new forms of matter appeared in our equations before they appeared in our laboratories. In fact, it's become the usual case. Quarks (both as a general concept, and the specific *c*, *b*, and *t* flavor varieties), color gluons, *W* and *Z* bosons, and all three kinds of neutrinos were seen first as solutions of equations, and only later as physical realities.

The search is on for other talented inhabitants of c-world that we'd like to recruit into p-world, notably Higgs particles and axions. It breaks my heart not to describe them in detail here, but that would require two major digressions, just as we're building to a climax. You can find more information and references on them in the glossary and endnotes and, for the Higgs particle, also in Appendix B.

For our story, the most important proposed expansion of the equations of physics is supersymmetry, often fondly called SUSY. Supersymmetry, as the name suggests, proposes that we should use equations with larger symmetry.

SUSY's new symmetry is related to the boost symmetry of special relativity. As you may recall, boost symmetry says that the basic equations don't change when you impart a common, constant velocity to all the components of the system you're describing. (Dirac had to modify Schrödinger's equation to give it that property.) Supersymmetry likewise says that the equations don't change when you impart a common motion to all the components of the system you're describing. But it's a very different kind of motion from what's involved in boost symmetry. Instead of motion through ordinary space with a constant velocity, supersymmetry involves motion into new dimensions!

Before you get carried away with visions of spirit worlds and wormholes through hyperspace, let me hasten to add that the new dimensions have a very different character from the familiar dimensions of space and time. They're *quantum* dimensions.

What happens to a body when it moves in the quantum dimensions isn't that it gets displaced—there's no notion of distance out there—instead, its spin changes. The "superboosts" turn particles with a given amount of intrinsic spin into particles with a different amount of spin. Because the equations are supposed to stay the same, supersymmetry relates properties of particles with different spin. SUSY allows us to see them as the same particle, moving in different ways through the quantum dimensions of superspace.

You can visualize the quantum dimensions as new layers of the Grid. When a particle hops into these layers its spin changes, and so does its mass. Its charges—electric, color, and weak—stay the same.

SUSY might allow us to complete the work of unifying the Core. Unification of the different charges, using the symmetry *SO*(10), united all the gauge bosons into a common cluster, and all the quarks and leptons into a common cluster. But no ordinary symmetry is capable of joining those two clusters, for they describe particles with different spins. Supersymmetry is the best idea we have for connecting them.

Correcting the Correction

After expanding the equations of physics to include supersymmetry, we find that they have more solutions. Just as with Maxwell's and Dirac's equations, the new solutions represent new forms of matter—new kinds of fields and, as their excitations, new kinds of particles.

Supersymmetry requires us, roughly speaking, to *double* the number of fields we have in our equations. Alongside every field we know about, off in the quantum dimensions, is a new partner field describing activity in the new layer of Grid. The particles associated with these new fields have the same charges (of all kinds) as their known partners, but different masses and spins.

It may sound reckless and extravagant to infer a doubling of the world based on esthetic considerations.[2] Maybe it is. But Dirac's introduction of antimatter involved a similar doubling, and Maxwell's enlargement of the world of light from the visible band to the infinite expanse of the electromagnetic spectrum was even larger. Both were—at the time of their conception—essentially esthetic gambits. So physicists have learned to be bold. It is more blessed to ask forgiveness than permission. Thus endeth the apology; now back to business.

The new, partner particles must be heavier than their observed siblings, or else they'd already have been observed. But we'll assume that they're not *too* much heavier, and see where that leads.[3]

Fluctuations in these new fields permeate the Grid. They are new kinds of virtual particles, which contribute to the screening and antiscreening of strong, weak, and electromagnetic sources. To see our way to the basics, at short distances or high energy, we have to correct our vision, to remove the distorting effect of that bubbly medium. We tried to make such a correction before, in Chapter 18, without taking into account these possible new contributions. Now we must correct the corrections.

Figure 20.1 displays the result. With SUSY on board, it works.

Gravity Too

We can also bring gravity into the game. Gravity, as we've seen, starts out *ridiculously* feebler than the other forces. Looking at Figure 20.1, at the left-hand side, corresponding to the distances and energies we can access in practice, we see that the difference in power between the strong and electromagnetic forces is roughly a factor

---

2. Reinforced by a striking numerical success—description upcoming.

3. For more on the quantitative aspects, see the following chapter and the endnotes.

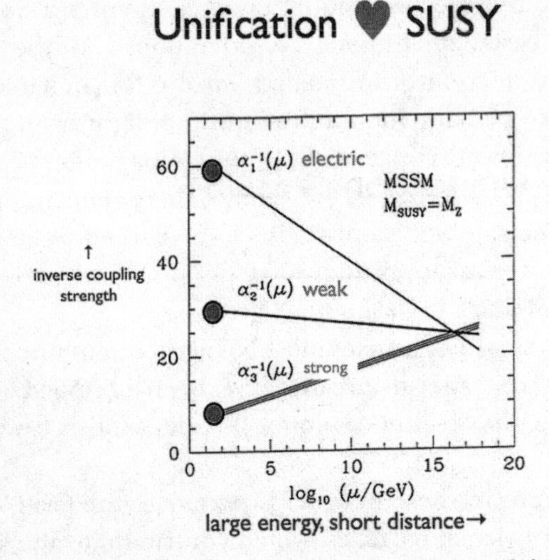

Figure 20.1 Supersymmetry requires expanding the equations of physics to include new fields. Grid fluctuations due to these new fields distort our view of the most basic, underlying physical process. After we correct for those distortions, we find accurate unification at short distances or, equivalently, at high energies.

of 10. So they fit easily, together with the weak interaction, into a nice tidy plot. Gravity doesn't come close to fitting in. Because it's feebler and we're plotting inverse power, gravity should appear up higher than the others. But to include it at this scale, we'd need to make our figure far larger than the known universe!

On the other hand . . .

For the Core forces—the strong, weak, and electromagnetic forces—the corrections as we go to the right, toward much shorter distances or higher energies, are fairly modest (and remember, each tick on the horizontal axis represents a factor of 10). Those corrections, after all, arose from a subtle quantum-mechanical effect: screening (or antiscreening) due to Grid fluctuations. When we look at gravity at *very* short distances, by transferring *very* large amounts of energy, the change is much more drastic. As we discussed way back in Chapter 3, gravity responds directly to

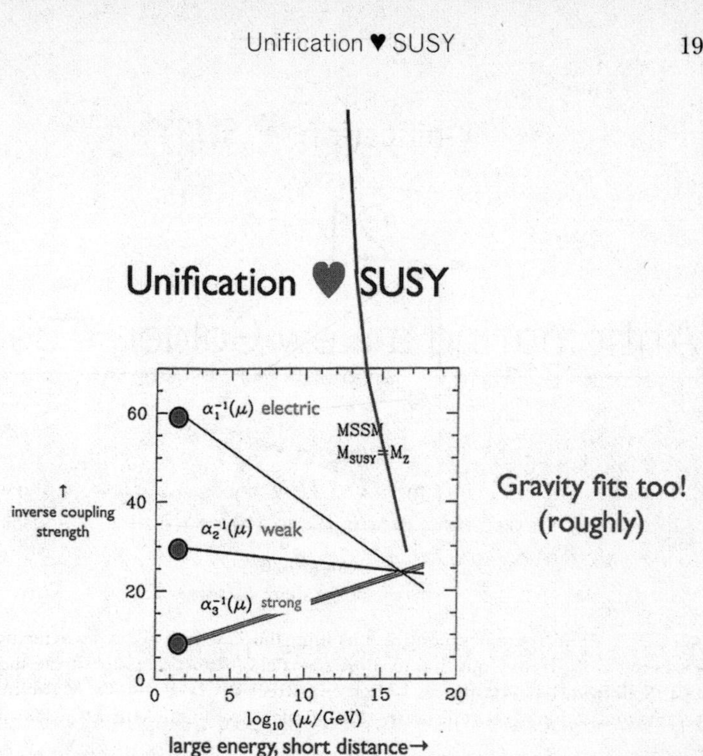

Figure 20.2 Gravity starts ridiculously feeble, but at short distances it approaches the other interactions in power; and they all come together, pretty nearly.

energy. Its power, as defined here, is proportional to energy squared. Allowing for that effect, we can calculate the power of gravity at short distances and compare it with the other interactions. Figure 20.2 displays the result. From well outside the known universe, the inverse power of gravity descends to join the other interactions, pretty nearly.

21

Anticipating a New Golden Age

We've made our case for unification. Now it goes out to Nature, the ultimate jury. We await verdicts from accelerators, from the cosmos, and from deep underground.

WE'VE SEEN THAT THE THEORIES of the Core forces, each deeply based on symmetry, can be combined. The three separate Core symmetries can be realized as parts of a single, all-encompassing symmetry. Moreover, that encompassing symmetry brings unity and coherence to the clusters of the Core. From a motley six, we assemble the faultless Charge Account. We also discover that once we correct for the distorting effect of Grid fluctuations—and after upping the ante to include SUSY—the different powers of the Core forces derive from a common value at short distances. Even gravity, that hopelessly feeble misfit, comes into the fold.

To reach this clear and lofty perspective, we made some hopeful leaps of imagination. We assumed that the Grid—the entity that in everyday life we consider empty space—is a multilayered, multicolored superconductor. We assumed that the world contains the extra quantum dimensions required to support supersymmetry. And we boldly took the laws of physics, supplemented with these two "super" assumptions, up to energies and down to distances far beyond where we've tested them directly.

From the intellectual success so far achieved—from the clarity and coherence of this vision of unification—we are tempted to believe that our assumptions correspond to reality. But in science, Mother Nature is the ultimate judge.

After the solar expedition of 1919 confirmed his prediction for the bending of light by the Sun, a reporter asked Albert Einstein what it would have meant if the result had been otherwise. He replied, "Then God would have missed a great opportunity." Nature doesn't miss such opportunities. I anticipate that Nature's verdicts in favor of our "super" ideas will inaugurate a new golden age in fundamental physics.

The LHC Project

Near Geneva, at the CERN laboratory, protons will race around a 27-kilometer circumference tunnel at 0.999998 times the speed of light. There will be two tight beams flowing in opposite directions. They will meet at four interaction points, where detectors the size of five-story office buildings will monitor the explosive output of collisions. This is the Large Hadron Collider (LHC) project. To get an impression of the size of the accelerator and a key detector, see Color Plates 8, 9, and 11.

In sheer size, the LHC is our civilization's answer to the pyramids of ancient Egypt. But it is a nobler monument in many ways. It is born out of curiosity, not superstition. It is a product of cooperation, not command.

And LHC's gigantic scale is not an end in itself, but a side effect of its function. Indeed the overall physical scale of the project is not its only, or even its most, impressive dimension. Inside the long tunnel are exquisitely machined and aligned superconducting magnets. Each of these powerful giants is 15 meters long, yet built to submillimeter tolerances. Precise timing is also essential, in the electronics. In separating the collisions and tracking the particles, nanoseconds count.

The flow of raw information that gushes out is not only mind-boggling but computer network-boggling. It is estimated that

LHC will produce 15 petabytes (1.5×10^{15} bytes) of information per year. This is equal to the bandwidth alloted half a million telephone conversations running all at once, nonstop. New architectures that allow thousands of computers around the world to divide the load are being developed to cope with it. This is the (computer) Grid project.

The LHC will achieve concentrations of energy large enough to test both of our two "super" assumptions.

We can make a pretty reliable estimate of what it takes to knock loose a piece of the condensate responsible for the Grid's (electroweak) superconductivity. The weak force is short-ranged, but not infinitely so. The *W* and *Z* bosons are heavy, but not infinitely so. The observed range of the force, and mass of the force carriers, give us good handles on the stiffness of the condensate responsible for those effects. Knowing the stiffness, we can estimate how much energy we need to concentrate in order to break off individual pieces (quanta) of the condensate—or, in more prosaic terms, to make the Higgs particle or particles or sector or whatever kind of new stuff it is that makes the Grid a cosmic superconductor. Unless our ideas are somehow *very* wrong, the LHC should be up to the job.

It's a similar story for supersymmetry. We want the Grid fluctuations associated with the new SUSY partner fields to bring the unification of coupling powers into line. If they're to do that job, then the excitations associated with those fields can't be too stiff. Some of their excitations—some new SUSY particles, partners of particles that we know—must be produced and detected at LHC.

If SUSY partner particles do show up, they'll give us new windows into the physics of unification. The masses and couplings of these particles, like the basic couplings of the Core interactions, will be distorted by effects of Grid fluctuations. But the details of the distortion are predicted to be different, in specific ways. If all goes well, the successful (but isolated and tenuously based) unification calculation we have today could blossom into a thriving ecology of mutually supportive results.

Dark Matter in the Balance

Toward the end of the twentieth century, physicists consolidated their extremely successful theory of matter: the Core. The Core's compact yet remarkably complete and accurate account of the basic laws of matter crowned centuries of work.

But just as that was happening, astronomers made startling new discoveries that restored our humility. They discovered that the matter we've been dealing with all those centuries—the kind of matter we study in biology, chemistry, engineering, and geology, the kind of matter we're made from, the kind we understand profoundly with our Core theory[1]—that stuff, *normal* matter, contributes only about 5% of the mass of the universe as a whole!

The remaining 95% contains at least two components, called dark energy and dark matter.

Dark energy contributes about 70% of the mass. It has been observed only through its gravitational influence on the motion of normal matter. It has not been observed to emit or absorb light; it is not dark in the usual sense, but transparent. Dark energy seems to be uniformly distributed throughout space, with a density that is also constant in time. The theory of dark energy is in bad shape. It's a problem for the future.

Dark matter contributes about 25% of the mass. It too has been observed only through its gravitational influence on the motion of normal matter. Dark matter is *not* uniformly distributed in space, nor is its density constant in time. It clumps, though not as tightly as normal matter. Around every galaxy that's been carefully studied, astronomers have found an extended halo of dark matter. These haloes are diffuse—their density is typically a million times less than that of normal matter, in regions where they overlap—but they extend over far larger volumes than normal matter. Rather than speaking of galaxies as objects with haloes, it

---

1. The kind of matter made from photons, electrons, quarks, and gluons.

might be more appropriate to speak of the normal-matter galaxy as an impurity in the dark matter.

The dark-matter problem, I think, is ripe for solution.

Among the new partner particles predicted by SUSY, one is special: the lightest. Its properties depend on details that we don't have convincing ideas about (especially, the specific values of all the SUSY partner masses). So we have to try all the possibilities. We find that in many cases the lightest SUSY partner is extremely long-lived (longer than the lifetime of the universe) and interacts very weakly with normal matter. The most striking thing, though, is that when we run our equations through the big bang to see how much of this stuff would have survived to the present day, we find that it's roughly the right amount to account for the dark matter. Naturally, all this suggests that the lightest SUSY partner *is* the dark matter.

So it's quite possible that by investigating the basic laws of physics at ultrashort distances, we'll solve a major cosmological riddle, and begin to shed some of that irksome humility. If a candidate particle to provide the dark matter does emerge, it will be a great enterprise to check whether that candidate really does the job. On the theory side, we'll need to pin down all the reactions relevant to its production in the big bang and run the numbers. On the experimental side, we'll want to check that the candidate really is what's out there. Once you know exactly what you're looking for, it becomes much easier to find it.

There's another promising idea about what the dark matter is, which emerges from a different proposal for improving the equations of physics. As we've discussed, QCD is in a profound and literal sense constructed as the embodiment of symmetry. There is an almost perfect match between the observed properties of quarks and gluons and the most general properties allowed by local color symmetry, in the framework of special relativity and quantum mechanics. The only exception is that the established symmetries of QCD fail to forbid one sort of behavior that is not observed to occur. The established symmetries permit a sort of interaction among gluons that would spoil the symmetry of the equations of QCD under a change in the direction of time. Experiments provide severe limits on the possible strength of that interaction. The limits are much more severe than might be expected to arise accidentally.

The Core theory does not explain this "coincidence." Roberto Peccei and Helen Quinn found a way to expand the equations that would explain it. Steven Weinberg and I, independently, showed that the expanded equations predict the existence of new, very light, very weakly interacting particles called axions. Axions are also serious candidates to provide the cosmological dark matter. In principle they might be observed in a variety of ways. Though none is easy, the hunt is on.

It's also possible that both ideas are right, and both kinds of particles contribute to the total amount of dark matter. Wouldn't that be pretty?

One Shoe Heard From; Listening for Others

Unification of the Core forces brings in a larger symmetry, and the larger symmetry brings in additional forces. We postulate a second, stiffer layer of cosmic superconductivity in the Grid to explain how the additional forces, which have not been observed, are suppressed.[2] But we don't want to suppress them completely. At the unification scale—at high energy or, equivalently, short distance—and beyond, these new interactions are united with the Core and have the same power.

Quantum fluctuations—virtual particles—that reach those extraordinary energies are extremely rare, but they do occur. Correspondingly, the effects that those fluctuations catalyze are predicted to be very small, but not zero. Two of these effects are so

---

2. For a deeper discussion of these matters, see Appendix B.

unusual and otherwise unexpected that they're regarded as classic signs of unification physics.

- Neutrinos should acquire mass.
- Protons should decay.

We've heard the first of those shoes drop. As mentioned before, neutrinos do have very small, but nonzero, masses. The observed values of those masses are broadly consistent with expectations from unification.

We're waiting for the other shoe. Deep underground, giant photon collectors monitor huge vats of purified water, looking for flashes that will signal the death of protons. Our estimates for the rate suggest that discovery should not be far off. If so, it will open yet another entry into unification physics—perhaps the most direct and powerful. For protons can decay in many ways, and the rates for the different possibilities directly reflect the new interactions arising from unification.

Unification of our theories of the Core interactions—strong, weak, and electromagnetic—into a single unified theory involves some guesswork, but the principles are clear. Quantum mechanics, special relativity, and (local) symmetry fit together smoothly. Using them, we can make definite suggestions for experimental exploration, including quantitative estimates of what to expect.

Unification with gravity also looks good, at the level of comparing the fundamental strength of all the interactions, as we've seen. But with gravity our ideas about what the unified theory is are nowhere near as concrete. The ferment of ideas around superstring theory seems promising, but no one's been able to pull these ideas together enough to suggest specifically what new effects to expect. What shoes will unification including gravity drop? Any that we can hope to hear? That too is a question for the future.

Epilogue

A Smooth Pebble, a Pretty Shell

> [T]o myself I seem to have been only a boy playing on the sea-shore, diverting myself in now and then finding a smoother pebble or a prettier shell than ordinary, whilst the great ocean of truth lay all undiscovered before me. —*Isaac Newton*

HAVING SCALED OUR THIRD PEAK, we've reached a natural stopping point. It's time to come to rest, look back, and scan the scenery.

Looking down on the valley of everyday reality, we perceive much more than before. Beneath the familiar, sober appearances of enduring matter in empty space, our minds envision the dance of intricate patterns within a pervasive, ever-present, effervescent medium. We perceive that mass, the very quality that renders matter sluggish and controllable, derives from the energy of quarks and gluons ever moving at the speed of light, compelled to huddle together to shield one another from the buffeting of that medium. Our substance is the hum of a strange music, a mathematical music more precise and complex than a Bach fugue, the Music of the Grid.

Through patchy clouds, off in the distance, we seem to glimpse a mathematical Paradise, where the elements that build reality shed their dross. Correcting for the distortions of our everyday vision,we create in our minds a vision of what they might really be: pure and ideal, symmetric, equal, and perfect.

Or have our imaginations made too much of a wispy chimera? We point our telescope, and wait for the clouds to clear.

Owning Up on Mass

Ahead lie other mountains, whose peaks we can't yet discern.

As promised, I've accounted for 95% of the mass of normal matter from the energy of massless building blocks, using and delivering on the promise of Einstein's second law, $m = E/c^2$. Now it's time for me to own up to what I've failed to explain.

The mass of the electron, although it contributes much less than 1% of the total mass of normal matter, is indispensable. The value of that mass determines the size of atoms. If you doubled the electron's mass, all atoms would contract to half their size; if you halved the electron's mass, all atoms would expand to twice their size. Other things would happen, too, that could make life in anything like the form we know it impossible. If electrons were heavier by a factor of 4 or so, then it would be favorable for electrons to combine with protons to make neutrons, emitting neutrinos. That would be R.I.P. for chemistry, to say nothing of biology, because neither electrically charged nuclei nor electrons would be available to build atoms and complex molecules.

And yet we have no good idea (yet) about why electrons weigh what they do. There's no evidence that electrons have internal structure (and a lot of evidence against it), so the kind of explanation we arrived at for the proton, relating its mass to internal energy, won't work. We need some new ideas. At present, the best we can do is to accommodate the electron's mass as a parameter in our equations—a parameter we can't express in terms of anything more basic.

It's a similar story for the masses of our friends the up and down quarks, u and d. They make a quantitatively small but qualitatively crucial contribution to the masses of protons and neutrons, and hence of normal matter. If their values were significantly different, life might become difficult or impossible. Yet we can't explain why they have the values they do.

We also don't understand the values of the masses of the electron's heavier, unstable clones—that is, the muon (μ) and tau (τ) leptons, which are, respectively, 209 and 3478 times heavier than electrons. We don't know where those numbers 209 and 3478 come from. Ditto for the heavier, unstable clones of the up quark—that is the charm (c) and top (t) quarks—and for the heavier, unstable clones of the down quark, strange (s) and bottom (b).

The only good news in this debacle is that all these quarks and leptons appear to be closely related to one another, both in terms of their observed properties and theoretically, within the unified theories we discussed in previous chapters. So if we manage to understand one—and if the unified theories are right!—it'll teach us about all the others.

The fact that we remain so ignorant about the origin of quark masses means that my explanation for the feebleness of gravity, based on the proton's being light compared to the Planck mass, isn't complete. I took it for granted that most of the mass of the proton arises from the energy of the quarks and gluons it contains, according to Einstein's second law. And that is actually true in Nature: the u and d quarks do in fact have only tiny masses, much smaller than the mass of the proton, so their direct contribution to the proton's mass is very small. But if you ask me *why* those quark masses are tiny, I don't have a solid answer (though I could spin out some tales).

Then there's the Higgs particle, sometimes said to be "the origin of mass" or even "the God particle." In Appendix B, I've sketched a wonderful circle of ideas centered around the Higgs particle. In brief, the Higgs field (which is more fundamental than the particle) enables us to implement our vision of a universal cosmic superconductor and embodies the beautiful concept of spontaneous symmetry breaking. These ideas are deep, strange, glorious, and very probably true. But they don't explain the origin of mass—let alone the origin of God. Although it's accurate to say that the Higgs field allows us to *reconcile* the existence of certain kinds of mass with details of how the weak interactions work, that's a far cry from explaining the origin of mass or why the different masses have the values they do. And as we've seen, most of the mass of normal matter has an origin that has nothing whatsoever to do with Higgs particles.

We also don't really understand the masses of neutrinos. And we *really* don't understand the masses of a welter of particles that appear in our theories but haven't yet been observed, including the Higgs particle or particles, all the particles associated with supersymmetry, axions, ...

A shorter way to summarize the situation would be to say that the only case in which we *do* understand the origin of mass is the one I've told you about in this book. Happily, that understanding covers the lion's share of the mass of normal matter—the matter made of electrons, photons, quarks, and gluons—the kind of matter that dominates our immediate environment, that we study in biology and chemistry, and that we're made of.

Darkness Revisited

It was a great discovery of astronomy (perhaps its greatest) that distant stars and nebulae are made from exactly the same sort of matter as we find on Earth. In recent decades, however, astronomers have partly undiscovered that basic truth. They've found that the bulk of the mass of the Universe, about 95%, comes from something else. New forms of matter, not made from electrons, photons, quarks, and gluons, are responsible for 95% of the mass of the universe.

The new stuff comes in at least two varieties, called dark matter and dark energy. Those aren't very good names, because one of the few things we know about this stuff is that it isn't dark: it doesn't absorb light to any detectable extent. Nor has it been observed to emit light. It appears perfectly transparent. Nor has it been observed to emit protons, electrons, neutrinos, or cosmic rays of any kind. In short, both the dark matter and the dark energy interact with ordinary matter only very feebly, if at all. The only way they have been detected is through their gravitational influence on the orbits of ordinary stars and galaxies, the things we do see.

We know very little for sure about dark matter. It might be made from supersymmetric particles, as I discussed before, or from axions. (I'm very fond of axions, in part because I got to name them. I used that opportunity to fulfill a dream of my youth. I'd noticed that there was a brand of laundry detergent called "Axion," which sounded to me like the name of a particle. So when theory produced a hypothetical particle that *cleaned up* a problem with an *axial* current, I sensed a cosmic convergence. The problem was to get it past the notoriously conservative editors of *Physical Review Letters*. I told them about the axial current, but not the detergent. It worked.) Heroic experiments are under way to test these possibilities and others, and with any luck we'll have much clearer ideas about what the dark matter is within a few years.

We know even less about dark energy. It seems to be spread out perfectly evenly, with the same density everywhere and everywhen, as if it were an intrinsic property of space-time. Unlike any conventional kind of matter (even supersymmetric particles or axions), the dark energy exerts negative pressure. It tries to pull you apart! Fortunately, although dark energy supplies about 70% of the mass of the universe as a whole, its density is only about 7×10^{-30} times the density of water, and its negative pressure cancels only about 7×10^{-14} of normal atmospheric pressure—less than a part in a trillion. I don't know when we'll have clearer ideas about what the dark energy is. I'd guess not very soon. I hope I'm wrong.

Acknowledgments

THIS BOOK CAME IN LARGE PART out of the public lectures "The Universe Is a Strange Place," "The World's Numerical Recipe," "The Origin of Mass and the Feebleness of Gravity," and "The Persistence of Ether" that I've given over the last few years at many institutions. I'd like to thank my hosts for the opportunity to give the lectures, and the audiences for many interesting questions and useful feedback.

I'd like to thank MIT for continuous support; Nordita for hospitality through the bulk of the writing; and Oxford University for hospitality during the completion of this book.

I'd like to thank Betsy Devine and Al Shapere for close readings of the manuscript that resulted in many important improvements. I'd also like to thank Carol Breen for her comments on the manuscript, and especially for her help with an early version of Chapter 6.

I'd like to thank John Brockman and Katinka Matson for urging me to write the book, and Bill Frucht, Sandra Beris, and the people at Perseus for lots of help and encouragement.

Betsy's support and input have been an essential inspiration to me throughout.

Appendix A

Particles Have Mass, the World Has Energy

AS WE DISCUSSED IN CHAPTER 3, $E = mc^2$ holds for isolated bodies at rest. For moving bodies, the correct mass-energy equation is

$$E = \frac{mc^2}{\sqrt{1 - \frac{v^2}{c^2}}}$$

where v is the velocity. For a body at rest ($v = 0$), this becomes $E = mc^2$.

When a body—for example, a proton or an electron—is accelerated, v generally changes, but m stays the same. Therefore, the equation tells us, E changes.

At first hearing, that might sound like just the opposite of what we discussed in the main text. We said that *energy* is conserved, but *mass* is not. What gives?

Conservation of energy applies to systems, not to individual bodies. The total energy of a system of bodies includes contributions from both energy of motion (given by the formula above) and "potential energy" terms that reflect the interactions among the bodies. The potential energy terms are given by other formulas, which depend on the distances between the bodies, their electric charges, and perhaps other things. It is only the total energy that is conserved.

An isolated body has a constant velocity. That's Newton's first law of motion, which, unlike his zeroth law, still appears to be valid. When a body is isolated, we can regard it as a system unto itself. So the energy of the body should be conserved—and, from the formula, it is.

Conversely, when a body's velocity changes, that very change is a signal that the body is not isolated. Some other body has to be acting on it to produce the change in velocity. The action of one body on another generally transfers energy between them. Only the total energy is conserved, not the energy of each body separately.

When we make a proton out of quarks and gluons, these concepts come together. From a fundamental perspective, a proton at rest is a complicated system of interacting quarks and gluons. The quarks and gluons individually have very little mass. That doesn't, however, prevent the whole system from having energy. Call that energy *E*. It's conserved in time, as long as the whole system—that is, the proton—is isolated. Alternatively, we can consider the isolated proton as a black box: a "body" with mass *m*. These two quantities, which arise in the alternative descriptions, are related by $E = mc^2$ (or $m = E/c^2$).

In Chapter 2, we considered a dramatic violation of conservation of mass. An electron and a positron annihilate, and out come a collection of particles whose total mass is 30,000 times larger. Nevertheless, energy is conserved. The velocities of the initial electron and positron were very close to the speed of light. Therefore, according to the general mass-energy equation, their energy is very large—much larger than mc^2. The particles that emerge from the collision, although they are more massive, move a bit more slowly. When you add up their energies, calculated using the mass-energy equation, the sum matches the total energy of the original electron and positron. (Once the particles fly out and separate, the interaction, or potential, energy becomes negligibly small.)

Finally, to complete this discussion of the relation between mass and energy, we must consider the special case of particles with zero mass. Important examples include photons, color gluons, and gravitons. Such particles move at the speed of light. If we attempt to put $m = 0$ and $v = c$ into our general mass-equation equation, both the numerator and the denominator on the right-hand side vanish, and we get the nonsensical relation $E = 0/0$. The correct result is that the energy of a photon can take any value. Photons of different energy differ neither in their velocity, which is always the speed of light c, nor of course in their mass, which always vanishes, but in their frequency (that is, the rate at which the underlying electric and magnetic fields oscillate). The energy E of a photon is proportional to the frequency ν of the light it represents. More precisely, they are related by the Planck-Einstein-Schrödinger equation $E = h\nu$, where h is Planck's constant.

For photons in the visible range, we sense that distinction as a difference in color: photons at the red end of the spectrum have the smallest energy, photons at the blue end the largest. Going down in energy, beyond the visible we meet infrared, microwave, and radio wave photons. Moving up, we meet ultraviolet, x-rays, and γ rays.

Appendix B

The Multilayered, Multicolored Cosmic Superconductor

We live inside an exotic superconductor that hides the symmetry of the world.

THE MOST FUNDAMENTAL PROPERTY of superconductors isn't that they conduct electricity extremely well (although they do). The most fundamental property was discovered by Walther Meissner and Robert Ochsenfeld in 1933. It is called the Meissner effect. What Meissner and Ochsenfeld discovered is that magnetic fields cannot penetrate into the interiors of superconductors, but are confined to a thin surface layer. Superconductors can't abide magnetic fields. That is their most fundamental property.

Superconductors get their name from a more obvious and spectacular property, their special talent for sustaining electric currents. Superconductors can carry electric currents that flow without resistance and therefore persist indefinitely, even without a battery to drive them. Here's the connection between the Meissner effect and such super(b) conductivity:

If we expose a superconducting body to an external magnetic field, then according to the Meissner effect, that body must find a way to cancel the magnetic field, so that there's no net field inside. The body can only ensure such cancellation by generating an equal and opposite magnetic field of its own. But magnetic fields are generated by currents. To generate a magnetic field that keeps the net field zeroed, the superconducting body must be able to support currents that persist indefinitely.

Thus the possibility of "super" flow of electric current follows from the Meissner effect. The Meissner effect is more fundamental. It is the true mark of a superconductor.

The Meissner effect applies not only to real magnetic fields but also to those that arise as quantum fluctuations. Thus the properties of virtual photons, which are fluctuations in electric and magnetic fields, get modified inside a superconductor. The superconductor does its best to cancel out those fluctuating magnetic fields. As a consequence, virtual photons inside a superconductor are rarer and extend over smaller distances than in empty[1] space.

In the Grid view of the world, electric and magnetic forces result from the interplay between charged sources and virtual photons, a.k.a. field fluctuations. Particle A affects the field fluctuations around it, which also affect another particle, B. This is our most fundamental picture of how a force arises between A and B. It is what you see depicted in the basic Feynman diagram, Figure 7.4.

So the fact that field fluctuations inside a superconductor become rare and short-ranged means that the corresponding electric and magnetic forces are effectively enfeebled. Especially, those forces cease to operate over long distances.

The field-nullifying supercurrents also make life harder for real photons inside superconductors. It takes more energy to make self-renewing field disturbances, which, we've learned, is what photons are. In the equations, this effect shows up as a nonzero mass for photons. In short: inside superconductors, photons are heavy.

Cosmic Superconductivity: Electroweak Layer

The weak interaction is a short-range force. The fields responsible for this force, the *W* and *Z*, are similar in many ways to the electromagnetic field. The particles that arise as disturbances in these

1. That is, "empty."

fields (the *W* and *Z* particles) resemble photons. Like photons, they are bosons. Like photons, they respond to charges—not electric charges to be sure, but what we've called green and purple charges, with similar physical properties. Their most obvious difference from photons is that *W* and *Z* are heavy particles. (Each weighs about as much as one hundred protons.)

Short-range force. Heavy particles. Sound familiar? It should. Those are exactly the properties of electromagnetic forces and photons inside superconductors.

The modern theory of electroweak interactions is heavily invested in the analogy between what happens to photons inside superconductors and the observed properties of *W* and *Z* bosons in the cosmos. According to this part of the Core theory, the entity we perceive as empty space—the Grid—*is* a superconductor.

Even though the conceptual and mathematical parallels run very deep, Grid superconductivity differs from conventional superconductivity in four main ways:

Occurrence Conventional superconductivity requires special materials and low temperatures. Even the new "high-temperature" superconductors max out at less than 200 Kelvin (room temperature is about 300 Kelvin).

Grid superconductivity is everywhere, and has never been observed to break down. Theoretically, it should persist up to about 10^{16} Kelvin.

Scale The photon mass inside a conventional superconductor is 10^{-11} proton masses, or less.

The *W* and *Z* masses are about 10^2 proton masses.

What Flows The supercurrents of conventional superconductivity are flows of electric charge. They cause electromagnetic fields to become short-range, and photons to acquire mass.

The supercurrents of Grid superconductivity are correlated flows of much less familiar types of charge: purple weak charge and hypercharge. *W* and *Z* fields can be generated by those flows, so the forces that *W* and *Z* generate become short-range, and the *W* and *Z* particles acquire mass.

Substrate Although many details remain mysterious, we understand in broad terms how conventional superconductors work. (For many superconducting materials, we have quite a detailed and accurate theory; for others, including the so-called high-temperature superconductors, it's a work in progress.) Specifically, we know what their superflows are made from. The supercurrents are flows of electrons, organized into what are called Cooper pairs.

By contrast, we don't have a reliable theory for what the Grid superflows are made from. None of the fields we've observed to date has the right properties. Theoretically, it's possible that the job is done by a single new field, the so-called Higgs field, and its attendant Higgs particle. It's also possible that several fields are involved. In the theories featuring SUSY, which figured heavily in our ideas for achieving unification, there are at least two fields contributing to the superflows, and at least five particles associated with them. (In the language of Chapter 8 there are two condensates, and five distinctive kinds of field disturbances.) Things could be even more complicated. We don't know. A major goal of the LHC project is to address these questions experimentally.

Grid superconductivity doesn't involve strong color charges, so the strong color gluons remain unsuppressed, with zero mass. Nor does it affect photons. Unlike *W* and *Z*, which have their influence largely suppressed and rendered short-range by field-canceling supercurrents, photons remain massless. Fortunately for us—given that our electrical and electronic technology, not to mention our chemical being, relies on vigorous electromagnetic forces—Grid supercurrents are electrically neutral.

Cosmic Superconductivity: Strongweak Layer

We can take these ideas a very important step further.

The central achievement of Grid superconductivity, for the Core electroweak theory, is to explain why the weak force *appears* much weaker and more obscure than the electromagnetic force, even though they appear on very nearly the same footing in the fundamental equations. (Indeed, as we've discussed, the weak force is, fundamentally, slightly more powerful.) In terms of the Core symmetries, it shows us how to explain the reduction

$$SU(3) \times SU(2) \times U(1) \rightarrow SU(3) \times U(1)$$

from the fundamental Core symmetries (strong × weak × hypercharge) to the ones that have long-range consequences (strong × electromagnetic).

In our unified theories we work with much larger symmetry groups than the Core $SU(3) \times SU(2) \times U(1)$, such as $SO(10)$. With more symmetry, you have more possibilities for transformations among the different kinds of charges, and more kinds of gluon/photon/*W,Z*-like gauge particles that implement those transformations.

The additional gauge particles are capable of doing things that rarely, if ever, happen in reality. By transforming, for example, a unit of weak color charge into a unit of strong color charge, we can change a quark into a lepton, or an antiquark. The Charge Account is full of such possibilities. So we can easily generate, for example, the decay

$$p \rightarrow e^+ + \gamma$$

of protons into positrons and photons. If this decay occurred with anything like a typical weak interaction rate, the decay would occur within a small fraction of a second. We'd be in serious trouble, because our bodies would quickly vaporize into electron-positron plasma.

We can suppress the unwanted processes, while retaining the underlying unification symmetry, using a new layer of Grid superconductivity. Then we'll have, as we proceed from very small to longer distances, the active (unsuppressed) fields reduced according to

$$SO(10) \rightarrow SU(3) \times SU(2) \times U(1) \rightarrow SU(3) \times U(1)$$

The second step is what we already had, in the Core.

For the first step, we need much more efficient Grid supercurrents. They must mightily suppress the unwanted strong↔weak color charge transformations. Of course, this means that the supercurrents themselves would be flows involving both strong and weak color charges.

No known form of matter can supply such supercurrents. On the other hand, it's easy to invent new Higgs-like fields that do the job. People have played with other ideas, too. Maybe these currents arise from particles racing around in additional, tiny curled-up spatial dimensions. Maybe they're vibrations of strings that wrap around additional tiny curled-up spatial dimensions. Because the concentrated energies required to probe distances this small lie far, far beyond what we can achieve in practice, these speculations are not easy to check.

Fortunately, just as in the Core electroweak theory, we can make good progress by taking the supercurrents as given, without *fingo*-ing hypotheses about what they're made of. That's the philosophy I adopted in Part III of this book. It led us to some encouraging successes, and to some specific predictions. If it survives further scrutiny, we'll be able to assert with confidence that we live within a multilayered, multicolored cosmic superconductor.

Appendix C

From "Not Wrong" to (Maybe) Right

SAVAS DIMOPOULOS IS ALWAYS ENTHUSIASTIC about something, and in the spring of 1981 it was supersymmetry. He was visiting the new Institute for Theoretical Physics in Santa Barbara, which I had recently joined. We hit it off immediately—he was bursting with wild ideas, and I liked to stretch my mind by trying to take some of them seriously.

Supersymmetry was (and is) a beautiful mathematical idea. The problem with applying supersymmetry is that it is too good for this world. We simply do not find particles of the sort it predicts. We do not, for example, see particles with the same charge and mass as electrons, but a different amount of spin.

However, symmetry principles that might help to unify fundamental physics are hard to come by, so theoretical physicists do not give up on them easily. Based on previous experience with other forms of symmetry, we have developed a fallback strategy, called spontaneous symmetry breaking. In this approach, we postulate that the fundamental equations of physics have the symmetry, but the stable solutions of these equations do not. The classic example of this phenomenon occurs in an ordinary magnet. In the basic equations that describe the physics of a lump of iron, any direction is equivalent to any other, but the lump becomes a magnet with some definite north-seeking pole.

A familiar, simple example of spontaneous symmetry breaking is the convention for driving on one side of the road. It doesn't matter which side of the road people drive on, as long as everybody does the same thing. If some people drive on the left and others on the right, it's an unstable situation. So the symmetry between left and right must be broken. Of course in different universes, call them U.S. and U.K., the choice may differ.

Surveying the possibilities opened by spontaneously broken supersymmetry requires model building—the creative activity of proposing candidate equations and analyzing their consequences. Building models with spontaneously broken supersymmetry that are consistent with everything else we know about physics is a difficult business. Even if you manage to get the symmetry to break, the extra particles are still there (just heavier) and cause various mischief. I briefly tried my hand at model building when supersymmetry was first developed in the mid-1970s, but after some simple attempts failed miserably, I gave up.

Savas was a much more naturally gifted model-builder, in two crucial respects: he did not insist on simplicity, and he did not give up. When I identified a particular difficulty (let us call it A) that was not addressed in his model du jour, he would say, "It's not a real problem, I'm sure I can solve it." And the next afternoon he would come in with a more elaborate model that solved difficulty A. But then we would discuss difficulty B, and he would solve that one with a completely different complicated model. To solve both A and B, you had to join the two models, and so on to difficulty C, and soon things got incredibly complicated. Working through the details, we would find some flaw. Then the next day Savas would come in, very excited and happy, with an even more complicated model that fixed yesterday's flaw. Eventually we eliminated all flaws, using the method of proof by exhaustion—anyone, including us, who tried to analyze the model would get exhausted before they understood it well enough to find the flaws.

When I tried to write up our work for publication, there was a certain feeling of unreality and embarrassment about the complexity and arbitrariness of what we had come up with. Savas was undaunted. He even maintained that some existing ideas about unification using gauge symmetry, which to me seemed genuinely fruitful, were not really so elegant if you tried to be completely realistic and work them out in detail. In fact, he had been talking to another colleague, Stuart Raby, about trying to improve those models by adding supersymmetry! I was extremely skeptical about this "improvement," because I was certain that the added complexity of supersymmetry would spoil the hard-won success of gauge symmetry in explaining the relative values of the strong, electromagnetic, and weak coupling constants. The three of us decided to do the calculation, to see how bad the situation was. To get oriented and make a definite calculation, we started by doing the crudest thing, which was to ignore the whole problem of breaking supersymmetry. This allowed us to use very simple (but manifestly unrealistic) models.

The result was amazing, at least to me. The supersymmetric versions of the gauge symmetry models, although they were vastly different from the originals, gave very nearly the same answer for the couplings.

That was the turning point. We put aside the "not wrong" complicated models with spontaneous supersymmetry breaking and wrote a short paper that, taken literally (with unbroken supersymmetry), was wrong. But it presented a result that was so straightforward and successful that it made the idea of putting unification and supersymmetry together seem (maybe) right. We put off the problem of how supersymmetry gets broken. And today, although there are some good ideas about it, there is still no generally accepted solution.

After our initial work, more precise measurements of the couplings made it possible to distinguish between the predictions of models with and without supersymmetry. The models with supersymmetry work much better. We all eagerly await operation of the Large Hadron Collider at CERN, the European particle physics laboratory, where, if these ideas are correct, the new particles of supersymmetry—or, you might say, the new dimensions of superspace—must make their appearance.

Glossary

Acceleration Rate of change of velocity. Acceleration is therefore the rate of change of the rate of change in position. Newton's central discovery in mechanics was that the laws governing accelerations are often simple.

Amplitude (quantum) Quantum mechanics supplies predictions for the probability of various events, but the equations of quantum mechanics are formulated using *amplitudes*, which are a sort of pre-probability. More precisely, the probability is the square of the amplitude. (For the cogniscienti: amplitudes are generally complex numbers, and the probabilities are the square of their magnitudes.) The term *amplitude* is used to describe the heights of waves of many kinds, such as ocean waves, sound waves, and radio waves. Quantum-mechanical amplitudes are essentially the heights of quantum-mechanical wave functions. For further discussion and examples, see Chapter 9. [See also: Wave function.]

Antimatter The matter we ordinarily encounter—and are made of—is based on electrons, quarks, photons, and gluons. Matter made from the corresponding antiparticles, namely antielectrons (a.k.a. positrons), antiquarks, photons, and gluons, is often called antimatter. (Note that photons and gluons are their own antiparticles. More accurately, some of the gluons are antiparticles of other ones; all eight gluons make a complete set, which goes over into itself when you take the antiparticle of everything.) [See also: Antiparticle.]

Antiparticle The antiparticle of a given particle has the same mass and spin as that particle, but the opposite value of electric charge and other conserved quantities. The first antiparticles discovered, historically, were antielectrons, also known as positrons. They were predicted theoretically by Dirac, and subsequently observed by Carl Anderson in cosmic rays. A deep consequence of quantum field theory is that for every kind of particle there is a corresponding antiparticle. The photon is its own antiparticle; this is possible because photons are electrically neutral.

Particle-antiparticle pairs can have zero values of all conserved quantum numbers; thus they can be produced from pure energy, and can also arise spontaneously as quantum fluctuations (virtual pairs).

Antiscreening The opposite of screening. [See: Screening.] Whereas screening diminishes the effective strength of a given electric charge, antiscreening enhances the effect of a color charge. Thus antiscreening allows a feeble seed color charge to become powerful far away. Antiscreening of color charge is the essence of asymptotic freedom, a key feature of QCD. [See also: Color charge, QCD.]

Asymptotic freedom The concept that strong interaction becomes weaker at short distances. More specifically, effective color charges, which govern the power of the strong interaction, become smaller at short distances. Reading it the other way, the power of a given, isolated color charge is enhanced far away. Physically, this happens because the source charge induces a cloud of virtual particles that enhances or *antiscreens* it. A consequence of asymptotic freedom is that radiation by a fast-moving color charge that moves in the same direction ("soft" radiation) is common, whereas radiation that changes the overall direction of flow is rare. The soft radiation supplies quarks with mates they can join up with to form hadrons; yet the overall flow follows the pattern set by the underlying quarks (and antiquarks and gluons). So we get to "see" quarks and gluons not as individual particles, but through the jets they trigger. We can eat our quarks and have them too. [See also: Charge Without Charge, Jets.]

Axion Hypothetical particle predicted in a class of theories that fix an esthetic flaw of the Core theory (for the record: the strong P, T problem). Axions are predicted to interact very weakly with ordinary matter, and to have been produced during the big bang with roughly the right density to provide the dark matter. Therefore axions are a well-motivated dark-matter candidate.

Baryon One of the two basic body plans for strongly interacting particles (hadrons). Baryons can be considered, roughly, as being formed from three quarks. More accurately, they result from letting three quarks come into equilibrium with the Grid. The complete wave function for a baryon contains, in addition to three quarks, arbitrary numbers of quark-antiquark pairs and gluons. Protons and neutrons, the building blocks of atomic nuclei, are baryons. [See also: Hadron.]

Boost A transformation that causes a system, including all its components, to move at a constant velocity. A modern perspective on special relativity is that it postulates *boost symmetry*. Thus the laws of physics are supposed to look the same after the application of a boost. As a consequence, there's no way to measure, by studying physical behavior purely within a closed, isolated system, how fast it's moving.

Boson Quantum theory, and more particularly quantum field theory, brings sharp new meaning to the concept of two objects being absolutely the same, or indistinguishable. If one has, say, two photons in states A and B at a given time, and two photons in states A', B' at a later time, one cannot say whether the transition involved was A → A', B → B' or A → B', B → A'. One must consider both possibilities. For bosons one adds their amplitudes, for fermions one subtracts. Photons are bosons. A consequence is that photons like to enter the same state, for then the transition amplitudes are doubled. Lasers exploit this effect. Along with photons themselves, gluons, *W*, and *Z* are bosons, as are mesons and the hypothetical Higgs particle. We often say that bosons obey Bose statistics, or Bose-Einstein statistics, after pioneer physicists who clarified the implications of this behavior for systems containing many identical particles.

Charge In electrodynamics, charge is the physical attribute to which electric and magnetic fields respond. (Magnetic fields respond only to moving charges.) In the quantum version, QED, we can simply say that charge is the thing that photons care about. Charge can be either positive or negative. Particles carrying electric charge of the same sign (either both positive or both negative) repel, whereas particles carrying electric charges of opposite sign attract. An important property of charge is that it's conserved. Each kind of fundamental particle carries an amount of charge—possibly zero—that's a stable characteristic of that kind of particle. For example all electrons have the same quantity of charge, usually denoted $-e$. (Confusingly, some authors use e, without the minus sign. There's no agreed-upon convention, as far as I know.) Protons have charge e, the opposite of electrons. The total charge of a system is simply the sum of the charges of everything in it. Thus atoms that contain an equal number of protons and electrons have zero charge overall. In the theory of the strong interaction, three additional kinds of charges, called color charges or simply color, play a central role. Color charges have properties similar to electric charge; for example, they are conserved. In QCD, color charge is what gluons care about. [See also: Field, Electrodynamics, Color, Chromodynamics.]

Charge Account Whimsical name for a unified account of quarks and leptons that fully explains the pattern of their strong (color), weak, and electromagnetic charges. The mathematical structure involved is the spinor representation of $SO(10)$, together with a specific choice of an $SU(3) \times SU(2) \times U(1)$ subgroup. The choice of subgroup specifies how the Core theory sits within the unified theory. If we allow only the transformations of the established, smaller Core symmetry, the unified Charge Account breaks into six disconnected pieces, and the peculiar pattern of electric charges (or, equivalently, hypercharges) is not explained. [See also: Unified theory.]

Charge Without Charge A conceptual consequence of asymptotic freedom. The effective color charge of a given source decreases at short distances. A nonzero, finite value of the charge at nonzero distance corresponds to zero charge in the mathematical limit of zero distance. Thus a point source generates Charge Without Charge. It's a feat worthy of the Cheshire cat.

Chromodynamics Theory that describes the activity of color gluon fields, including their response to color charges and currents (flows of charge). It is the accepted theory of the strong force. Mathematically, chromodynamics is a generalization of electrodynamics. Because quantum theory is important in all applications of chromodynamics, it is often referred to as quantum chromodynamics, or QCD for short. [See also: Strong force.]

Color 1. A fundamental physical attribute, analogous to electric charge, but different. There are three kinds of color charge, here called red, white, and blue. A quark carries a unit of one of these color charges. Gluons carry both a positive unit and a negative unit of color charge, possibly for different colors. 2. In everyday life, of course, color means something completely different. Namely, color is the frequency of electromagnetic radiation, when that frequency falls within a narrow band that corresponds to the peak radiation from our Sun. That's a joke, sort of. Actually, everyday use of *color* is prescientific. It refers to our eyes' and brains' response to such electromagnetic radiation. [See also: Charge, Chromodynamics.]

Confinement The fact that quarks are never observed in isolation. More precisely, for any observable state, the number of quarks minus the number of antiquarks is a multiple of 3. (Note that 0 is a multiple of 3.) Confinement is a mathematical consequence of chromodynamics, but not one that is easy to demonstrate.

Conservation law A quantity is conserved if its value, for an isolated system, does not change with time. Charge, energy, and momentum are important examples of conserved quantities. Conservation laws are extremely important, for they provide stable landmarks amid the incessant flux of quantum Grid.

Core Our working theory of strong, electromagnetic, weak, and gravitational interactions. It is based on quantum mechanics, three sorts of local symmetry—specifically, the transformation groups $SU(3)$, $SU(2)$, $U(1)$—and general relativity. The Core theory gives precise equations governing all elementary processes that are presently known to occur. Its predictions have been tested in many experiments and proved to be accurate. The Core theory contains esthetic flaws, so we hope it is not Nature's last word. (In fact it can't be, because it doesn't describe the dark matter.)

Cosmological term A logical extension of the equations of general relativity. In geometric language, the cosmological term encourages (depending on its sign) either the uniform expansion or contraction of space-time. Alternatively, the cosmological term can be interpreted as representing the influence of a constant density of energy (positive or negative) on metric field. That density ρ is accompanied by a pressure p related to it by the "well-tempered equation" $\rho = -p/c^2$.

Dark energy Astronomical observations indicate that a large portion of the mass of the Universe, around 70% of the total, is uniformly distributed and very accurately transparent. Other, independent observations indicate an accelerating expansion of the universe, which we can ascribe to negative pressure. The magnitudes and relative sign of these effects are consistent with the well-tempered equation. Thus the observations, so far, can be described by a cosmological term. It is logically possible, however, that future observations will reveal that the density and/or pressure are not constant, or that they are not related by the well-tempered equation. The term dark energy was introduced to avoid prejudging these issues.

Dark matter Astronomical observations indicate that a large portion of the mass of the universe, around 25% of the total, is distributed much more diffusely than ordinary matter and is very accurately transparent. Galaxies of ordinary matter are surrounded by extended halos of dark matter. The halo weighs, in total, about five times as much as the visible galaxy. There may also be independent condensations of dark matter. Interesting candidates to provide dark matter include weakly interacting massive particles (WIMPS), associated with supersymmetry, and axions. [See also: Supersymmetry, Axions.]

Dirac equation Invented by Paul Dirac in 1928, it is a modification of Schrödinger's equation for the quantum-mechanical wave function of electrons, designed to be consistent with boost symmetry (that is, special relativity). The Dirac equation is, roughly speaking, four times as big as the Schrödinger equation. More precisely, it is a set of four interconnected equations, governing four wave functions. The four components of the Dirac equation automatically incorporate spin (up or down) for both particles and antiparticles, which accounts for four components. With minor modifications the Dirac equation is also used to describe quarks and neutrinos. In today's physics we interpret the Dirac equation as an equation for the field that creates and destroys electrons (or, equivalently, destroys and creates positrons), rather than as an equation for the wave function of individual particles.

Electrodynamics Theory that describes the activity of electric and magnetic fields, including their response to charges and currents (flows of charge). It can also be considered the theory of the photon field. Light in all its forms, including (for instance) radio waves and x-rays, is now understood to be activity in electric and magnetic fields. The basic equations for electrodynamics were discovered by Maxwell and perfected by Lorentz. [See also: Charge, Maxwell equations.]

Electron A fundamental constituent of matter. Electrons carry all the negative electric charge in normal matter. They occupy the bulk of the space in atoms, outside the atoms' small nuclei. Electrons are very light and mobile compared to nuclei, so they are the active players in most of chemistry and of course in electronics.

Electroweak theory The modern theory governing both the weak interactions and electromagnetic interactions. It is sometimes also called the standard model. There are two leading ideas in the electroweak theory. One is that its equations are governed by local symmetry, leading to the equations of Maxwell and Yang-Mills. The other is that space is an exotic kind of superconductor, which—roughly speaking—shorts out some of the interactions, hiding their effects. (Another important idea is that the interactions are chiral. This is rather more technical, and I won't try to describe it here. Its most dramatic consequence is that the weak interactions violate parity—that is, the symmetry between left and right.) It is sometimes said that the electroweak theory unifies QED and the weak interactions, but it would be more accurate to say that it mixes them. [See also: Weak force.]

Energy A central concept in physics. Given its importance, it is surprising how subtle and unpromising the definition of energy at first appears. Indeed, it was only in the mid-nineteenth century that the modern concept of energy and its conservation began to emerge. The original and most obvious form of energy is kinetic energy, which is energy related to motion. (In pre-relativistic mechanics the kinetic energy of a body was identified as one-half its mass times the square of its velocity. The relativistic formulas, which include rest energy, are discussed in Appendix A.) The kinetic energy of a body generally changes when forces act upon it, but for certain kinds of forces (so-called conservative forces) it is possible to define a potential energy function, depending only on the position, such that the total of kinetic and potential energies is constant. More generally, for systems of bodies, and a certain class of forces, the sum of all their kinetic energies, plus a potential energy depending on their positions, is conserved. The first law of thermodynamics asserts that energy is conserved, although it can be hidden as heat, which is a manifestation of very small-scale, hard-to-observe motion within bodies. In effect, the first law of thermodynamics asserts that the fundamental forces of nature will always be found to be conservative. That bold hypothesis, put forward long before the nature of the fundamental forces was at all clear, is justified by the success of thermodynamics. [See also: Jesuit Credo.]

In modern theories of physics, energy appears as a primary concept, on the same footing as time, to which it is deeply related. For example, the time T it takes for the oscillating electric disturbances within a photon to go through one complete cycle is related to the photon's energy E through $ET = h$, where h is Planck's constant. Within these theories, conservation of energy follows from symmetry of the equations under time translation—or, in common English, the fact that the laws don't change over time.

You might wonder: if the fundamental laws of physics ensure conservation of energy, why are people urged to take measures to conserve energy? After all, the laws of physics are supposed to be self-enforcing! The point is that some forms of energy are more useful than others; in particular, random jiggling (heat) is not readily available to do useful work. It would be better physics to ask people to minimize their entropy production. [See also: Entropy.]

Entropy A measure of disorder. [See a book on thermodynamics, or consult Wikipedia.]

Ether A space-filling material. Before physicists got comfortable with the concept of fields as fundamental components of reality, they tried to make mechanical models of electric and magnetic fields. They speculated that electric and magnetic fields described the arrangements of more basic particle-like objects, much as the density and flow-fields of liquids describe arrangements and rearrangements of atoms. Those models got complicated and never worked very well, so that the concept "ether" got a bad reputation. In modern physics, however, a space-filling medium is the primary reality. This medium has very different properties from the classical ether, so I've given it a new name: the Grid.

Fermion Quantum theory, and more particularly quantum field theory, brings sharp new meaning to the concept of two objects being absolutely the same, or indistinguishable. If one has, say, two electrons in states A and B at a given time, and two electrons in states A', B' at a later time, one cannot say whether the transition involved was A → A', B → B' or A → B', B → A'. One must consider both possibilities. For bosons one adds their amplitudes, for fermions one subtracts. Electrons are fermions. A consequence is the Pauli exclusion principle: two electrons cannot get into the same state, for then the amplitudes cancel completely. Pauli exclusion induces an effective repulsion between electrons (quantum statistical repulsion), which is responsible for the fact that different electrons in atoms must occupy different states, which in turn is largely responsible for the fact that chemistry is a rich and complicated subject. Not only electrons, but all leptons and quarks, and their antiparticles, are fermions. So are protons and neutrons, which is a big reason why *nuclear* chemistry is a rich and complicated subject. We often say that fermions obey Fermi statistics, or Fermi-Dirac statistics, after pioneer physicists who clarified the implications of this behavior for systems containing many identical particles.

Feynman graph (also called Feynman diagram) Feynman graphs are a pictorial shorthand for processes described by quantum field theory. They consist of lines connected at hubs (also called vertices). The lines represent free motion of particles through space-time; the hubs represent interactions. With this interpretation, the Feynman graph depicts a possible process in space-time: some (real or virtual) particles interact, and their quantum state may change as a result. There are precise rules for assigning probability amplitudes to the process depicted in a given Feynman graph. The square of the amplitude, according to the rules of quantum theory, is the probability of the process.

Field A space-filling entity. The field concept entered physics in the nineteenth century, through the work of Faraday and Maxwell on electricity and magnetism. They discovered that the laws of electricity and magnetism could be formulated most easily if one introduced the idea that there are (invisible) electric and magnetic fields filling all space. The force felt by an electric charge at a point gives a measure of the strength of the electric field at that point; but in the Faraday-Maxwell concept, the field is there, whether or not there is a charged particle around to sense it. Thus fields have a life of their own. The fruitfulness of this concept soon emerged, when Maxwell discovered that self-renewing disturbances in electric and magnetic fields could be interpreted as light, with no reference to material charge or currents whatsoever.

In the quantum version of electrodynamics, electromagnetic fields create and destroy photons. More generally, the kinds of excitations we sense as particles (electrons, quarks, gluons, etc.) are created and destroyed by various fields (electron fields, etc.), which are the primary objects. This provides our most fundamental understanding of the vital fact that any two electrons, anywhere in the universe, have exactly the same basic properties. Both were made by the same field!

Sometimes physicists or engineers will make statements like "I've reduced the electric and magnetic fields to zero inside my specially shielded laboratory." Don't be fooled! What this means is that the *average value* of those fields has been zeroed; nevertheless, the electromagnetic field, as an entity, is still present. In particular the electromagnetic field will still respond to charge currents inside the shielding, and it still boils with quantum fluctuations—that is, virtual photons. Similarly, the average values of electric and magnetic fields in deep outer space are zero, or nearly so, but the fields themselves extend throughout, and support the propagation of light rays over arbitrarily large distances. (The field destroys a photon at one point, and a new one is created at the next point, ...) [See also: Quantum field.]

Fitzgerald-Lorentz contraction The effect that structures within a moving body appear to a stationary observer as contracted (shrunk) in the direction of motion. Fitzgerald and Lorentz postulated such an effect in order to explain some observations in the electrodynamics of moving bodies. Einstein showed that the Fitzgerald-Lorentz contraction is a logical consequence of boost symmetry in the form suggested by Maxwell's equations—that is, of special relativity.

Flavor In today's physics, a poorly understood three-valued attribute of quarks and leptons, independent of their charges. For example, there are three distinct flavors of *U* quarks—*u* (up), *c* (charm), and *t* (top). Each has the same electric charge, $2e/3$, and one unit of color charge (red, white, or blue). In addition, there are three flavors of *D* quarks—*d* (down), *s* (strange), and b (bottom), also each in three colors, with electric charge $-e/3$. Similarly, there are three flavors of leptons—*e* (electron), μ (muon), and τ (tau lepton)—that have electric charge $-e$ and no color charge, and finally, three kinds of neutrinos with neither electric nor color charge. Within each of these groups, the particles of different flavor have identical Core interactions. They differ in mass, sometimes by large factors (for example, *t* is at least 35,000 times heavier than *u*). The weak interactions allow transformations among the different flavors. There is no good theoretical explanation of why the masses are what they are.

Although *W* bosons act to change flavors, it would be wrong to think that flavor plays the same role in weak interactions as color plays in the strong interaction. W bosons do not respond directly to the flavor property, but to yet another pair of charges, what I've called weak color charges. The *W* bosons change flavors for sport, so to speak; it is not what drives them. The triplication of particles with identical arrays of charges and the rules of the flavor-changing game that *W* bosons play remain mysteries.

Force 1. In Newtonian physics, force is defined to be an influence that, when acting on a body, causes it to accelerate. This definition was fruitful, and remains useful, because in many cases forces turn out to be simple. For example, an isolated body feels no forces—a statement equivalent to Newton's first law of motion, which says that an isolated body moves with constant velocity. 2. In modern fundamental physics—and, specifically, in the Core theory and its unified extensions—the old force concept is less useful. It is still common to use expressions such as "the strong *force*," "the weak *force*," and so on, but among themselves physicists usually refer, more abstractly, to "the strong *interaction*." I've generally used the one syllable version in this book.

Gauge symmetry [See: Local Symmetry.]

General relativity Einstein's theory of gravity, based on the idea of curved space-time or, alternatively, the metric field. In the field formulation, general relativity broadly resembles electromagnetism. But whereas electromagnetism is based on the response of electric and magnetic fields to charges and currents, general relativity is based on the response of metric field to energy and momentum. [See also: Metric field.]

Gluon Any of a set of eight particles that mediate the strong force. Gluons have properties similar to those of photons, but they respond to (and change) color charges rather than electric charge. The equations for gluons have tremendous local symmetry, which largely determines their form. [See also: Chromodynamics, Local Symmetry, Yang-Mills equations.]

Grid The entity we perceive as empty space. Our deepest physical theories reveal it to be highly structured; indeed, it appears as the primary ingredient of reality. Chapter 8 is all about the Grid.

Hadron A physical particle based on quarks and gluons. (Quarks and gluons themselves don't count, because they can't exist in isolation.) Two basic kinds of hadrons have been observed: mesons and baryons. Mesons are made by letting a quark and an anti-quark come into equilibrium with the Grid. Baryons are made by letting three quarks come into equilibrium with the Grid. Many dozens of different mesons and baryons have been observed. Nearly all are highly unstable. "Glueballs," initiated by two or three gluons, may also exist. There is some controversy about whether glueballs have been observed—the particles we discover don't come clearly labeled!

Higgs particle Excitation in the (so far, hypothetical) field that makes empty space a cosmic superconductor for the weak force.

Hub A space-time point at which particles (real or virtual) interact. In Feynman graphs, hubs are the places where three or more lines meet. Theories of particle interactions supply rules for what kinds of hubs are possible, and the mathematical factors associated with them. In the technical literature, hubs are usually called vertices. [See also: Feynman graph.]

Hypercharge The average electric charge of several particles related by symmetry. In the context of unified theories, hypercharge is more fundamental than electric charge; the distinction, however, arises at a finer level of technical detail than I've attempted in most of this book.

Jesuit Credo "It is more blessed to ask forgiveness than permission." This is a profound truth.

Jet A clearly identifiable group of particles moving in nearly the same direction. Jets of particles are frequently observed as products of high-energy collisions at accelerators. Asymptotic freedom allows us to interpret jets as the visible manifestation of underlying quarks, antiquarks, and gluons.

LEP Acronym for Large Electron-Positron collider. LEP operated at the great European laboratory CERN, near Geneva, through the 1990s. Roughly speaking, it took pictures of empty space, at even higher resolutions than SLAC. To do this, electrons and their antiparticles (positrons) were accelerated to enormous energies and then made to annihilate, producing an intense flash of energy within an extremely small volume. LEP was a machine for creative destruction. Experiments at LEP tested and established the Core theory with extraordinary quantitative precision. [See also: SLAC.]

Lepton Any of the particles e (electron), μ (muon), and τ (tau lepton) or their neutrinos. These particles carry zero color charge. e, μ, and τ all carry the same electric charge, $-e$. (Yes, I realize that the same symbol is being used for different things. Sorry, but letters often are, if you think about it.) Neutrinos carry zero electric charge. They all participate in the weak interactions.

There are very good (but *not* perfect) conservation laws, according to which the total number of electrons-antielectrons plus the total number of electron neutrinos–electron antineutrinos stays constant in time (even though the separate numbers may change)—and similarly for μ and τ. For example, in muon decay the end product is an electron, a muon neutrino, and an electron antineutrino. Both the initial state and the final state have muonic lepton number one and electronic lepton number zero. These "laws of conservation of lepton number" are violated by the phenomenon of neutrino oscillations. Small violation of lepton number conservation was predicted by unified theories. Its observation encourages us to think that those theories are on the right track. [See also: Neutrino.]

LHC Acronym for Large Hadron Collider. The LHC occupies the old LEP tunnel at CERN. It uses protons instead of electrons and positrons, and attains higher energies. It will be surprising if great discoveries don't occur at the LHC. At a minimum, we should find out what makes the Grid an exotic superconductor.

Local symmetry (also known as gauge symmetry) Symmetry supporting independent transformations in different space-time regions. Local symmetry is a very powerfull requirement, which few equations satisfy. Conversely, by assuming local symmetry, we are led to very specific equations, of the Maxwell and Yang-Mills type. Exactly these equations characterize the Core theory and the world. Local symmetry is also called gauge symmetry, for interesting but obscure historical reasons. [See also: Symmetry, Maxwell equations, Yang-Mills equations.]

Mass A property of a particle or system, which is a measure of its inertia (that is, a particle's mass tells us how difficult it is to change the particle's velocity). For centuries scientists thought that mass is conserved, but now we know that it is not.

Mass Without Mass The concept, realized in modern physics, that objects with nonzero mass can emerge from building blocks of zero mass.

Maxwell equations The system of equations governing the behavior of electric and magnetic fields, including their response to electric charges and currents. Maxwell arrived at the full set of equations in 1864, by codifying all the mutual influences among electricity, magnetism, charge, and current known at the time, and in addition postulating a new effect, which made the system consistent with conservation of charge. Maxwell's original formulation was somewhat messy, so the underlying (profound) simplicity and symmetry of the equations were not evident. Later, physicists including (notably) Heaviside, Hertz, and Lorentz cleaned things up and gave us the Maxwell equations as we know them today. These equations have survived the quantum revolution intact, although the interpretation of electric and magnetic fields has evolved. [See also: Field.]

Meson A type of strongly interacting particle, or hadron. [See: Hadron.]

Metric field A field that can be regarded as specifying, at a space-time point, the units for measuring time and distance (in all directions). Thus space itself comes supplied with measuring rods and clocks. Ordinary rods and clocks translate this underlying structure into accessible forms. Matter affects the metric field, and vice versa. Their mutual interaction is described by the general theory of relativity and gives rise to the observed force of gravity. [See also: General relativity.]

Momentum A central concept in physics. The original and most obvious form of momentum is kinetic momentum, associated with the motion of particles. In pre-relativistic mechanics, the kinetic momentum of a body was identified as its mass times its velocity. Newton called momentum "quantity of motion," and it appears in his second law: the rate of change of momentum of a body is equal to the force acting on it. In special relativity, momentum is closely related to energy. Under boosts, energy and momentum mix with one another, just as time and space do. The total momentum of an isolated system is conserved.

In modern theories of physics, momentum appears as a primary concept, on the same footing as space, to which it is deeply related. For example, the distance D that it takes for the spatially periodic electric disturbances within a photon to go through one complete cycle is related to the photon's momentum P through $PD = h$, where h is Planck's constant. Within these theories, conservation of momentum follows from symmetry of the equations under spatial translation—or, in common English, from the fact that the laws are the same everywhere. [Compare: Energy.]

Neutrino A kind of elementary particle that has neither electric charge nor color charge. Neutrinos are spin ½ fermions. Neutrinos come in three separate types, or flavors, associated with the three flavors of charged leptons (electrons e, muons μ, and tau leptons τ). In weak interaction processes, charged leptons and their antiparticles can be transformed into neutrinos and their antiparticles, but always in such a way that the lepton numbers are conserved. [See also: Leptons.] Neutrinos are emitted abundantly by the Sun, but their interactions are so feeble that nearly all of them pass freely through the Sun, not to mention Earth if they happen to point our way. Nevertheless, some representatives of the small fraction that do interact have been detected, in heroic experiments. Recently it was established that the different types of neutrinos, as they travel over long distances, oscillate from one form into another (for example, an electron neutrino might morph into a muon neutrino). Such oscillations violate the lepton conservation laws. Their existence and rough magnitude is consistent with expectations from unified theories.

Neutron A readily identifiable combination of quarks and gluons, and an important component of ordinary matter. Individual neutrons are unstable; they decay into proton, electron, and electron antineutrino with a lifetime of about fifteen minutes. Neutrons bound into nuclei can be stable, however. [Compare: Proton.]

Normal matter The physical substance studied in biology, chemistry, materials science and engineering, and most of astrophysics. It is, of course, the substance humans, and their pets and machines, are made of. Normal matter is made from u and d quarks, electrons, gluons, and photons. We have a precise, accurate, and remarkably complete theory of normal matter: the kernel of the Core.

Nucleus Small central part of an atom, where all the positive charge and almost all the mass is concentrated.

Particle Localized disturbance in the Grid.

Photon Minimal excitation of the electromagnetic field. A photon is the smallest unit of light, sometimes called the quantum of light. (By the way, a quantum leap is a very *small* leap.)

Planck's constant A physical constant that plays a central role in quantum theory. It appears, for example, in the relationship $E = h\nu$ between the energy of a photon and the frequency ν of the light it represents, or the relationship $p = h/\lambda$ between the momentum of a photon and the wavelength λ of the light it represents.

Planck units Units of length, mass, and time derived from the values of quantities that appear in physical laws, rather than from reference objects. Thus one doesn't need a standardized "meter stick" (or royal extremity) to compare lengths, nor a spinning Earth to fix the unit of time, nor a prototype kilogram. Instead, Planck units are constructed by taking suitable powers and ratios of the speed of light c, Planck's constant h, and Newton's constant G that appears in the equations for gravity. Planck units are not used in practical work: the Planck length and time units are absurdly small; the Planck mass unit is absurdly large on atomic scales, but awkwardly small on human scales. Planck units are important theoretically, however. They pose the challenge of *computing* such purely numerical quantities as the proton mass in Planck units (whereas there is no possibility of computing the mass of a standard kilogram, so the "problem" of computing the proton mass in kilograms is poorly posed).

If we extrapolate the existing laws of physics far beyond where they've been tested, we find that quantum fluctuations in the metric field on length and time scales below the Planck length and time are so important, compared to its average value, that the operational meaning of length and time becomes unclear. As discussed in this book, there are significant indications that the different forces of nature become unified—in particular, that their powers become equal—when measured in Planck units at the Planck scale of distance.

Proton A very stable combination of quarks and gluons. Protons and neutrons were once thought to be fundamental particles; now we understand that they are complex objects. One can get a useful model of atomic nuclei using the idea that they are bound systems of protons and neutrons. Protons and neutrons have nearly the same mass; neutrons are about 0.2% heavier. Protons have electric charge *e*, equal in magnitude but opposite in sign to the charge of electrons. The nuclei of hydrogen atoms are single protons. Protons are known to have lifetimes of at least 10^{32} years (much longer than the lifetime of the universe), but unified theories predict that their lifetime cannot be much longer than the existing limit, and important experiments are under way to test this prediction.

QCD Short for quantum chromodynamics. [See also: Chromodynamics.]

QED Short for quantum electrodynamics. It is, of course, the version of electrodynamics incorporating quantum theory. The fields now have spontaneous activity (virtual photons), and their disturbances come in discrete, particle-like units (real photons). [See also: Electrodynamics, Photon, Quantum field.]

Quantum field A space-filling entity that obeys the laws of quantum theory. Quantum fields are the legitimate children of the marriage between quantum mechanics and special relativity. Quantum fields differ from classical fields in that they exhibit spontaneous activity, also known as quantum fluctuations or virtual particles, at all times and throughout space. The Core theory, which summarizes our best current understanding of fundamental processes, is formulated in terms of quantum fields. Particles appear as secondary consequences; they are localized disturbances in the primary entities—that is, in quantum fields.

Some general consequences of quantum field theory that do not follow from quantum mechanics or classical field theory separately are: the existence of particle types that are the same throughout the universe and for all time (for instance, all electrons have exactly the same properties); the existence of quantum statistics [See: Boson, Fermion]; the existence of antiparticles; the inevitable association of particles with forces (for example, from the existence of electric and magnetic forces one deduces photons); the ubiquity of particle transformation (quantum fields create and destroy particles); the necessity of simplicity and high symmetry for consistent laws of interaction; and asymptotic freedom [see: Asymptotic freedom]. All these consequences of quantum field theory are prominent aspects of physical reality as we find it.

Quarks Together with gluons, the players in the strong force (experimental aspect) or, alternatively, QCD (theoretical aspect). They are spin ½ fermions. There are three distinct flavors of *U* quarks—*u* (up), *c* (charm), and *t* (top)—and each has the same electric charge, $2e/3$, and one unit of color charge (red, white, or blue); in addition there are three flavors of *D* quarks—*d* (down), *s* (strange), and *b* (bottom), also each in three colors, with electric charge $-e/3$. Weak interaction processes can transform the various flavors into one another. Thus (color) gluons change the color charge of a quark, but not its flavor, whereas *W* bosons change flavor, but not color. Quarks are not observed directly, but leave their signature in jets (experimental aspect) and are used as building blocks to construct the observed hadrons (theoretical aspect). All the Core interactions conserve the total number of quarks minus antiquarks. This "conservation of baryon number" ensures proton stability. Unified theories generally predict the existence of interactions that change quarks into leptons and could cause protons to decay. So far, no such decay has been observed. [See also: All the underlined terms.]

RHIC Acronym for Relativistic Heavy Ion Collider. RHIC lives on Long Island, at the Brookhaven National Laboratory. Collisions at RHIC reproduce, in a very small volume and for very brief periods, extreme conditions the like of which were last seen in the earliest moments of the big bang.

Schrödinger equation An approximate equation for the wave function of an electron. The Schrödinger equation does not satisfy boost symmetry, which is to say it is not consistent with special relativity. But it gives a good description of electrons that are not moving too fast, and it is easier to work with than the more accurate Dirac equation. The Schrödinger equation is the foundation for most practical work in quantum chemistry and solid-state physics.

Screening A positive electric charge attracts negative charges, which tend to cancel (screen) it. Thus the full strength of a positive charge will be felt only close by; far away, its influence is diminished by the accumulated negative charge. Screening is very important in the theory of metals, which contain mobile electrons. It is also important for "empty space," i.e., the Grid. In that case, the negative charges are supplied by virtual particles. Although particular virtual particles come and go, the total population is steady and makes the Grid a dynamic medium. [See also: Antiscreening, Grid, Virtual particle.]

SLAC Acronym for Stanford Linear Accelerator Complex, a facility that played a key role in establishing the Core theory. Here Friedman, Kendall, Taylor, and their collaborators took their high-resolution, short-exposure pictures of proton interiors, which put us on the road to QCD. The two-mile-long electron accelerator they used provided, in effect, an ultrastroboscopic nanomicroscope.

Spin The spin of an elementary particle is a measure of its angular momentum. Angular momentum, in turn, is a conserved quantity that has much the same relation to rotations in space as (ordinary) momentum has to translations in space. [See also: Momentum.] In classical mechanics, the angular momentum of a body is a measure of the angular motion of that body. The spin of an elementary particle is either a whole number or a whole number + ½ times $h/2\pi$, where h is Planck's constant. The magnitude of spin is a stable characteristic of each particle type. Leptons and quarks are said to have spin ½, because their spin is ½ times $h/2\pi$. Protons and neutrons also have spin ½. Photons, gluons, and W and Z bosons have spin 1. π mesons and the hypothetical Higgs particle have spin 0. The polarization of light is a physical manifestation of the photon's spin.

The angular momentum of an isolated body is conserved. To change an angular momentum, one must apply a torque. Rapidly spinning gyroscopes have large angular momentum, and this fact is largely responsible for their unusual response to forces.

Spontaneous symmetry breaking When the stable solutions of a set of equations have less symmetry than the equations themselves, we say that the symmetry is spontaneously broken. Notably, this can happen if it is energetically favorable to form a condensate or background field, as discussed in Chapter 8 and Appendix B. The stable solution will then have space filled with material, whose properties change under some (former) symmetry transformations. Such a transformation is no longer a distinction without a difference—now it does make a difference! Its associated symmetry has been spontaneously broken.

Standard model Terminology designed to make one of the greatest intellectual achievements of humankind sound boring. It is sometimes used to mean the electroweak part of the Core theory, sometimes to include both the electroweak theory and QCD.

Strong force One of the four basic interactions. The strong force binds quarks and gluons into protons, neutrons, and other hadrons, and holds protons and neutrons together to form atomic nuclei. Most of what is observed at high-energy accelerators is the strong force in action.

Superconductor Some materials, when they are cooled to very low temperatures, transition into a phase where their response to electric and magnetic fields exhibits qualitatively new features. Their electric resistance vanishes, and they largely expel—that is, cancel out—applied magnetic fields (this is the Meissner effect). We say they have become superconductors. The electromagnetic behavior of superconductors, when analyzed mathematically, indicates that within superconductors photons have acquired nonzero mass.

Although they resemble photons in many ways—they are spin 1 bosons, they respond to (weak color) charges—*W* and *Z* bosons have nonzero mass. On the face of it, that nonzero mass would rule out the otherwise attractive idea that *W* and *Z* bosons, like photons, obey equations with local symmetry. Superconductivity shows the way forward. By postulating that space is a superconductor, not for electric charge, but for the charges that *W* and *Z* bosons care about, we give those particles nonzero mass, while maintaining the local symmetry of the underlying equations. This is a central idea of modern electroweak theory, and it seems to describe Nature very well. More speculative ideas postulate an additional layer of Grid superconductivity, to give very large masses to the particles that mediate quark-lepton transitions.

Superstring theory A collection of ideas for extending the laws of physics. It has inspired brilliant work by brilliant people, resulting in important applications to pure mathematics. At present, superstring theory does not provide equations that describe concrete phenomena in the natural world. Specifically, the Core theory, which accurately describes so much about the physical world, has not been shown to emerge as an approximation to superstring theory.

The ideas of superstring theory are not necessarily inconsistent with the Core theory, or with the ideas for unification advocated in this book. But the ideas discussed here did not arise, historically, from superstring theory, nor have they been derived from superstring theory. Their origin is, as I've explained at length, partly empirical and partly mathematical/esthetic.

Supersymmetry (SUSY) A new kind of symmetry. Supersymmetry makes transformations among particles that have the same charges, but different spins. In particular, it allows us to see bosons and fermions, despite their radically different physical properties, as different views of a single entity. Supersymmetry can be realized as a boost symmetry in superspace, an extension of space-time to include additional quantum dimensions.

Our present Core equations do not support supersymmetry, but it is possible to expand them so that they do. The new equations predict the existence of many new particles, none of which has been oberved. One must postulate some form of Grid superconductivity to make many of these particles heavy. The good news is that the new particles, in their virtual form, support a successful, quantitative unification of the forces, as described in Chapter 20. One of the new particles could be a good candidate to supply the dark matter. The LHC should be powerful enough to produce some of the new particles, if they exist.

Symmetry Symmetry exists when there are distinctions that don't make any difference. That is, an object—or a set of equations—exhibits symmetry when there are things you can do to it that *might* have changed it but in fact do not. Thus an equilateral triangle has symmetry under rotation by 120° around its center, whereas a lopsided triangle does not.

Time dilation The effect that, viewed from the outside, the flow of time within a moving system appears to slow down. Time dilation is a consequence of special relativity.

Unified theory The different components of the Core theory are based on common principles—quantum mechanics, relativity, and local symmetry—but within the Core theory they remain separate and distinct. One has separate symmetry transformations for the color charges of QCD, the weak color charges of the standard electroweak theory, and hypercharge. Under these transformations, the quarks and leptons fall into six unrelated classes (actually eighteen, counting flavor triplication). All this structure begs us to consider the possibility of a larger, encompassing symmetry. Upon exploring the mathematical possibilities, one finds that many things click into place quite nicely. A relatively small expansion of the equations enables one to view all the known symmetries as parts of a satisfying whole, and to bring the scattered quarks and leptons together. As a bonus, gravity, which seemed hopelessly feebler than the other fundamental forces, comes into focus as well. To make the ideas work in quantitative detail, it appears we must also include supersymmetry. The expanded equations predict the existence of many new particles and phenomena. As explained in Chapters 19–21, the jury is out on these theories, but some verdicts should be coming in soon.

Velocity Rate of change of position.

Vertex [See: Hub.]

Virtual particle Spontaneous fluctuation in a quantum field. Real particles are excitations in quantum fields that have some useful degree of permanence and can be observed. Virtual particles are transients, which appear in our equations but not in experimental detectors. By supplying energy, it is possible to amplify the spontaneous fluctuations above threshold, an effect that makes (what would have been) virtual particles into real ones.

Wave function In quantum theory, the state of a particle is not specified by a position, or a definite direction for spin; instead, the primary description of the state involves its wave function. The wave function specifies, for each possible position and direction of spin, a complex number, the so-called probability amplitude. The (absolute) square of the probability amplitude gives the probability of finding the particle at that position with that direction of spin. For systems of many particles, or fields, the wave function similarly specifies amplitudes for all possible physical behaviors that you might find, upon making a measurement. A simple, but not too simple, example of wave functions at work is discussed in Chapter 9.

Weak force (or interaction) Together with gravity, electromagnetism, and the strong force, one of the fundamental interactions of Nature. Also known as the weak interaction. The most important effect of the weak interaction is to support transformations among different types of quarks and different types of leptons (but not of quarks into leptons or vice versa. Those hypothetical quark-lepton transformations arise only in unified theories.). This is what drives many radioactive decays and some crucial reactions of stellar burning.

Well-tempered equation $\rho = -p/c^2$ [See also: Cosmological term, Dark energy.]

Yang-Mills equations Generalizations of Maxwell's equations, which support symmetry among several kinds of charges. Speaking roughly and colloquially, we can say that the Yang-Mills equations are the Maxwell equations on steroids. Today's theories of the strong and of the electroweak interactions are largely based on the Yang-Mills equations, for the symmetry groups $SU(3)$ and $SU(2) \times U(1)$, respectively.

Notes

Chapter 1

A short and very accessible, though now in parts dated, introduction to physics by a grandmaster is *The Character of Physical Law* by Richard Feynman (MIT). *The Feynman Lectures on Physics* (3 volumes) by Richard Feynman, Robert Leighton, and Matthew Sands (Addison-Wesley) was prepared with Caltech undergraduates in mind, but the early parts of each book and of many individual chapters are conceptual, colloquial, and frequently brilliant.

Chapter 2

11 **the monumental work that perfected classical mechanics:** The classic analysis of the foundations of classical mechanics is *The Science of Mechanics* by Ernst Mach (Open Court). Einstein read this book carefully in his student days, and its critical discussion of the Newtonian concepts of absolute space and time helped him toward the relativity concept. He wrote, "Even [James Clerk] Maxwell and [Heinrich] Hertz, who in retrospect appear as those who demolished the faith in mechanics as the final basis of all physical thinking, in their conscious thinking adhered throughout to mechanics as the secured basis of physics. It was Ernst Mach who, in his history of mechanics (*Geschichte der Mechanik*), shook this dogmatic faith. This book exercised a profound influence upon me in this regard while I was a student. I see Mach's greatness in his incorruptible skepticism and independence." For Newton's own views, in his own words, see especially *Newton's Philosophy of Nature* (Hafner). For other historical and philosophical perspectives, see *The Concept of Mass* by Max Jammer (Dover).

Chapter 3

The Principle of Relativity (Dover) is an indispensable collection of classic papers on relativity. It contains papers by Lorentz, Einstein, Minkowski, and Weyl. Both of Einstein's two original special relativity papers and his foundational paper on general relativity are included. The first half of Einstein's first special relativity paper is almost free of equations, and it's a joy to read. The early parts of his first presentation of general relativity are also accessible and inspiring. (For students of physics: that paper as a whole remains the best introduction to general relativity, in my opinion.) *The Evolution of Physics* by Einstein and Leopold Infeld (Touchstone) is a charming popularization not only of relativity itself but also of its intellectual background in electromagnetism and of the foundations of field physics. Two good modern introductions to relativity are *Spacetime Physics* by Edwin Taylor and John Wheeler (Freeman) and *It's About Time: Understanding Einstein's Relativity* by David Mermin (Princeton).

Chapter 4

23 **95% of the mass**: As we'll see, most of the mass of normal matter can be calculated within a theory whose building blocks are massless gluons and massless *u* and *d* quarks and nothing else (the theory I call QCD Lite). QCD Lite truly generates Mass Without Mass. It is, however, not a complete theory of nature. Lots of things are left out: electromagnetism, gravity, electrons, the small intrinsic masses of *u* and *d* quarks, and more. Fortunately, we can estimate how much the things we've left out or idealized away affect the mass of normal matter—and we can check our estimates, using the sorts of calculations described in Chapter 9. To make a long story short, the left-out effects change things by less than 5%. (For experts: the most important effect comes from the *s* quark. It is too heavy to treat as massless, but not so heavy that we can integrate it out cleanly.)

Chapter 5

Richard Rhodes's book *The Making of the Atomic Bomb* (Simon & Schuster) is not only a masterpiece of history and of literature, but also an excellent introduction to nuclear physics.

Chapter 6

33 **a class of equations . . . by Chen Ning Yang and Robert Mills:** *50 Years of Yang-Mills Theory*, edited by G. 't Hooft (World Scientific), is an important collection of articles by leading experts on the physics that's grown out of the Yang-Mills equations.

34 The claim **without having to supply samples or make any measurements** is a (remarkably slight) exaggeration. It would be true if all quarks had either zero mass or infinite mass. Finite, nonzero values of their mass can come only from measurements or samples. In Nature the *u* and *d* quarks have almost zero mass, relative to the mass of protons or neutrons, whereas the *c*, *b*, and *t* quarks are so

heavy that they play very little role in the structure of protons and neutrons, even as virtual particles. The strange quark *s* is intermediate: it plays some role in the structure of protons and neutrons, though not a large one. We can get a good approximate theory of protons and neutrons by pretending that the *u* and *d* quarks have zero mass, whereas the others have infinite mass, and can therefore be ignored. I call this approximate theory "QCD Lite." In QCD Lite, you really don't have to make any measurements or supply any samples.

Einstein emphasized the ideal of purely conceptual theories, which do not require measurements or samples as input, in his *Autobiographical Notes*: "I would like to state a theorem which at present can not be based upon anything more than upon a faith in the simplicity, i.e., intelligibility, of nature: there are no arbitrary constants . . . that is to say, nature is so constituted that it is possible logically to lay down such strongly determined laws that within these laws only rationally completely determined constants occur (not constants, therefore, whose numerical value could be changed without destroying the theory)." QCD Lite is a rare example of a powerful theory of that kind. (For experts: another example is the theory of structural chemistry based on Schrödinger's equation with infinitely heavy nuclei.) This issue is closely related to the question of fixing parameters that arises in Chapter 9, and to the philosophical/methodological discussions in Chapters 12 and 19.

Because quarks do not appear as isolated particles, the concept of their mass requires special consideration. For short times and distances, quarks move as if they are free (asymptotic freedom). We can calculate some consequences of such motion, which of course depend on what value we assign to the quarks' mass. Then, when we compare calculations to experiment, we determine the value of the mass. This works well for the heavier quarks. For the lighter quarks, a more practical way is to calculate the contribution of their mass to the mass of the hadrons that contain them, as described in Chapter 9. Intuitively, what we mean by a quark's mass is the mass of the *bare* quark, stripped of its cloud of virtual particles.

43 **strictly identical conditions** assumes that there are no hidden variables describing protons—that is, that their only degrees of freedom are position and direction of spin. All applications of Fermi statistics to protons rely on this assumption. Their success therefore provides overwhelming evidence for it.

45 **no internal structure** brings up a very interesting and important issue, which arises not only for quarks but also for protons, nuclei, atoms, and molecules. Let's discuss it for protons. As I mentioned in the preceding note, there is overwhelming evidence that the state of a proton is completely specified by its position and spin. Yet our best theory of protons constructs them as complicated systems of quarks and gluons, or, more accurately (looking ahead to Chapters 7 and 8), as complicated patterns of disturbance in the Grid. How does all that structure get hidden? If there's all that stuff rattling around inside, why can't different protons have a large variety of different states, depending on what exactly that stuff inside is doing?

In classical physics, there would be many possible internal states—or, if you like, many "hidden variables." But those states are removed by the Quantum Censor. In quantum theory (looking ahead, again, to Chapter 9) we learn that the proton—or any quantum system—supports all its possible internal states at once, with different probability amplitudes. To get the lowest-energy *quantum* state, the proton combines many *classical* states, each with an appropriate amplitude. The next-best quantum state has a completely different set of amplitudes and much higher energy. As a consequence, you have to disturb the proton *quite a lot* to change its internal structure *at all*. Small disturbances don't supply enough energy to rearrange the amplitudes. So for small disturbances there is always a unique set of amplitudes; variations are censored. The internal structure is, in effect, frozen out. It is similar to the way a snowball acts as a hard, rigid sphere, even though it's made from lots of molecules that, at higher temperatures, would flow as liquid water.

An even closer analogy, mathematically, is to the physics of musical instruments. If you play a flute properly, it will sound a definite, desired tone (depending on the fingering, of course). Only if you blow too hard, or erratically, will it start to sound overtones and screeches. The desired tone corresponds to a particular—in detail, quite complex—vibration pattern of air in the flute. The overtone corresponds to a distinctly different pattern. In the quantum theory we have vibrating wave functions instead of vibrating air, but the concepts and the mathematics are closely similar. Indeed, when the "new" quantum theory using wave functions was discovered, physicists went back to their texts on acoustics for mathematical guidance.

It is because of the Quantum Censor that seemingly radical ideas about the deep structure of matter can turn out to have little practical consequence. For example, it is widely speculated that quarks are secretly strings. Yet we have an accurate theory, QCD, which covers many precise experiments (so far, all of them), but takes no account of that possibility. How is this possible?

If a quark is secretly a string, then the quantum-mechanical wave function for the quark will support configurations of the underlying string with many different sizes and shapes, weighted by their amplitudes. As time progresses, those different configurations evolve into one another, but the overall distribution will remain constant.

As long as the distribution of string configuration amplitudes stays the same, it's inert and undetectable. And it might cost a lot of energy to change the distribution. The internal, string degrees of freedom are invisible to experiments that do not reach that critical energy. For practical purposes they might as well not exist. Nobody's sure what the critical energy for quark string vibrations is, but it must be considerably larger than what any existing accelerator has attained.

51 **the theory we call quantum chromodynamics, or QCD:** A lively historical account of the ideas and experiments leading to QCD is *The Hunting of the Quark* by Michael Riordan (Touchstone). Two good, accessible accounts of the physics of

QCD and the standard model of electroweak interactions are *The Theory of Almost Everything* by Robert Oerter (Pi Press) and *The New Cosmic Onion* by Frank Close (Taylor and Francis). Unique, and a must, is Feynman's presentation of QED from scratch: *QED: The Strange Theory of Light and Matter* (Princeton).

55 **soft radiation is common:** A more refined account of the distinction between soft and hard radiation is based on the connection between the momentum of a gluon and the wavelength of its wave function. Low momentum corresponds to long wavelength. Long waves do not resolve the fine structure of the quark's cloud, so they respond to the cloud as a whole, with its amplification of color charge through antiscreening. Short waves do resolve the internal structure. The ups and downs of these waves tend to cancel out their interaction with the cloud, leaving the contribution of the seed charge, which is now resolved.

Chapter 7

58 **symmetry is a word that's in common use:** A classic introduction to symmetry, by a great pioneer of the mathematics of the subject and a profoundly cultured human being, is Hermann Weyl's *Symmetry* (Princeton). Eugene Wigner brought group theory into modern physics in a big way, and his thoughtful essays in *Symmetries and Reflections* (Ox Bow) are interesting from many angles.

69 **Grooks:** Piet Hein's witty Grooks are collected at http://www.chat.carleton.ca/~tcstewar/grooks/grooks.html.

72 **textbooks:** The nitty-gritty of quantum field theory is not for sissies. If you'd like to go deeper into the subject, I'd recommend starting with Feynman's *QED*, mentioned earlier, and my review article *Quantum Field Theory* written for the 100th anniversary of the American Physical Society, reprinted in *More Things in Heaven and Earth*, ed. Bederson (Springer). It is also posted at itsfrombits.com. For some years the leading textbook has been *An Introduction to Quantum Field Theory* by Michael Peskin and Daniel Schroeder (Addison-Wesley); an excellent new contender is *Quantum Field Theory* by Mark Srednicki (Cambridge). Tony Zee's *Quantum Field Theory in a Nutshell* (Princeton) treats many unusual aspects of the subject in a breezy style. Finally, the trilogy *Quantum Theory of Fields* (Cambridge) by Steven Weinberg is a magisterial statement from a grand master, but aside from the historical introduction in Volume 1, nonprofessionals are likely to find it very hard going.

Chapter 8

Biography of Einstein: There are many biographies of Einstein. Two superb ones, which emphasize his science, are his own *Autobiographical Notes* in *Albert Ein-*

stein, Philosopher-Scientist, edited by P. Schilpp (Library of Living Philosophers), and *Subtle Is the Lord* by Abraham Pais (Oxford). Pais was an eminent physicist in his own right.

Biography of Feynman: Feynman didn't write a systematic autobiography, but his personality shines through in his collections of anecdotes *Surely You're Joking, Mr. Feynman!* (Norton) and *What Do You Care What Other People Think?* (Norton). *Genius* by James Gleick (Pantheon) is a well-written, deeply researched account of Feynman's colorful life.

78 **inconsistent:** Inconsistent with what? Conservation of charge. Maxwell applied the known equations to a "thought-circuit" including what we'd now call a capacitor and found they required electric charge to appear from nowhere. Because the experimental evidence for conservation of charge under all circumstances seemed very strong, Maxwell modified the equations accordingly.

85 **". . . searching in the dark . . .":** Quote from an address at the University of Glasgow, 1933. The "simultaneity" quote is from his *Autobiographical Notes.*

88 **inevitability of a field:** In this discussion of the necessity of fields, I refer to a universal value of "now," solving for the fields in the future in terms of the fields now, and so on. How can that be legitimate, given the demise of simultaneity?

Technical answer: In a boosted frame, the horizontal slice "now" will get changed into a tilted slice. But because the equations take the same form, it will still be possible to calculate the value of the fields off the slice, in terms of their values on the slice. (Strictly speaking, you have to know the value both of the fields and of their time derivatives.) In short: different "now," same argument.

Nevertheless there is an important tension here, which makes it difficult to marry quantum theory and relativity. In the equations of quantum theory, and in their interpretation, time appears in a very different way from space. But in the equations of relativity, time and space get mixed up. So when we're doing quantum mechanics we make a very strong *distinction* between time and space, but we have to show—if we believe in relativity—that in the end it doesn't make a *difference.* Fundamentally, this is why it's difficult to construct quantum theories that are consistent with special relativity. The only way we know how to do it uses the elaborate formalism of quantum field theory (or possibly the even more elaborate—and still incomplete—formalism of superstring theory). The flip side of the difficulty is that we're led to a very tight, specific framework: (for experts: local) quantum field theory. Thankfully, this turns out to be the framework that Nature uses in our Core theory of physics. Going back to the marriage metaphor: if you're picky about what you'll accept in a partner, then if you find one at all it's likely to be a good one!

94 **so-called weak interaction:** The books by Close and Olmert mentioned previously contain extensive discussions of the weak interaction.

97 **maps of the world:** *Lectures on Classical Differential Geometry* by Dirk Struik (Dover) contains a nice discussion of the mathematics of map making. The standard reference emphasizing the geometric approach to general relativity is *Gravitation* by Charles Misner, Kip Thorne, and John Wheeler (Freeman). A

classic presentation emphasizing the field approach is *Gravitation and Cosmology* by Steven Weinberg (Wiley). I should emphasize that there is no contradiction between these approaches, and good physicists keep both in mind.

102 **Superstring theory:** *The Elegant Universe* by Brian Greene (Vintage) is a popular, enthusiastic presentation of string theory.

102 **a promising opportunity:** Recent developments in cosmology increasingly suggest that the Universe underwent a period of very rapid expansion, known as *inflation*, early in its history. *The Inflationary Universe* by Alan Guth (Perseus) is an excellent popular account of the underlying theory, by its founding father. Among the things that get inflated, according to theory, are quantum fluctuations in the metric field. Those fluctuations, magnified to cosmological dimensions, might be detectable today. Ambitious experiments to look for this effect are planned.

The precise cause of inflation (if indeed it occurred) is unknown. But a likely culprit emerges from combining two of the ideas discussed in this chapter:

- We discussed how empty space is presently filled with various material condensates. At extremely high temperatures these condensates could "melt" or otherwise change their character. We say there is a phase transition, conceptually similar to the familiar phase transitions (solid) ice → (liquid) water → (gas) steam; but here we're talking cosmic phase transitions. Because space itself changes properties, in effect the laws of physics change.
- One of the things that changes, in such cosmic phase transitions, is the energy of the condensate. As we'll discuss very shortly, this change shows up as a contribution to the dark energy. Very plausibly, then, the very early universe might have contained a much higher density of dark energy than we see today. Today's dark energy is causing the expansion of the universe to accelerate, but only gently. A much larger density early on would have triggered much more rapid acceleration.

And that's how inflation might arise.

108 *The Extravagant Universe* by Mark Kirshner (Princeton) is a personal account of the observations by one of the leading astronomers.

110 **A popular speculation:** These ideas are clearly explained and advocated in *The Cosmic Landscape* by Leonard Susskind (Times Warner).

Chapter 9

113 **our classical computers:** These steps describe what's involved in solving the equations straightforwardly. There are clever tricks that sidestep some of them in special cases. They go by names like Euclidean field theory, Green's function Monte Carlo, stochastic evolution, and so forth. It's a very technical subject, way beyond the scope of this book. Progress in solving the equations

of quantum mechanics could change the world, by allowing us to replace experiments in chemistry and material science with calculations. Progress in computing aerodynamics has largely accomplished this program for aircraft design, so that new designs can be tried out numerically, bypassing rounds of prototyping and wind tunnel testing.

114 **quantum computers:** Two directions of spins—up or down—can be interpreted as 1s and 0s, so they can be interpreted as bits. But, as we'll discuss in detail over the next few pages, the quantum state of a set of spins can describe many arrangements of the spins simultaneously. Therefore, it's possible to imagine operating on many different bit configurations at the same time. It's a kind of parallel processing enabled by the laws of physics. Nature appears to be very good at it, solving the equations of quantum mechanics very quickly and without much apparent effort.

We're not so good, at least not yet. The problem is that the different spin configurations interact with the outside world in different ways, and that disrupts the orderly parallel processing we'd like to do. The challenge of building a quantum computer is to find ways of keeping spins from interacting with the outside world, or correcting for the interactions, or engineering objects less delicate than spins that obey similar equations. It's an active area of research; there's no design that's an obvious winner.

117 **the famous EPR paradox:** Sharper, quantitative forms of the EPR paradox, involving concepts such as Bell's inequality and Greenberger-Horne-Zeilinger states, are described in books on the foundations of quantum mechanics. A good clear one is *Consistent Quantum Theory* by Robert Griffiths (Cambridge). There is an enormous literature concerned with different interpretations of quantum theory, tests of its elementary ingredients, and so forth. IMHO, if you see a skyscraper standing tall and erect for decades, even under heavy bombardment, you should begin to suspect that its foundation is ultimately sound, even if it's not clearly visible. Then again, conservation of mass once looked very secure as well. . . .

118 **a thirty-two-dimensional world:** This note is strictly for experts. The unnormalized amplitudes describe a space of thirty-two *complex* dimensions. This corresponds to sixty-four real dimensions. In normalizing the state, we lose two of these. So really we're dealing with a space of sixty-two dimensions.

120 **highly infinite:** The quantum continuum is so complicated to construct that it's tempting to think that we should somehow get rid of it. Edward Fredkin and Steven Wolfram are prominent advocates of this view.

Crude attempts certainly don't work. Without entering into debates, I'll just say—without fear of contradiction—that nothing remotely approaching the completeness, precision, and accuracy of the Core theory has emerged from any significantly different rival ideas. On the other hand, it is disconcerting to have limiting processes (and therefore calculations that are, in principle, infinitely long) appearing in the most basic formulation of physical laws.

But do they, really? It's not clear to me that genuine infinities arise if we only ask the theory to answer questions that we can also pose experimentally. Experimentally, we have a limited amount of time and energy available, and we can only make measurements of limited precision. And *approximate* calculations don't require that we actually take the limit!

This note is making me dizzy, so I'd better end it now.

122 **the errors will be small:** I'd like to devote this short and skippable paragraph to a very important, though slightly technical, conceptual point. You might worry about the errors introduced replacing continuous space-time by a discrete lattice. In many scientific problems, such as predicting the weather or modeling climate, that's a huge issue. But here, thanks to asymptotic freedom, it's not so bad. Because the quarks and gluons interact weakly at short distances, you can calculate analytically—that is, with pen and paper—the effect of replacing the true activities by their local averages, tracked at the lattice points. Then you can correct for it.

124 **theory is not . . . predicting . . . just accommodating:** The mass m_π of the pion is the most sensitive to m_{light}; the mass m_K of the K-meson K is the most sensitive to m_s, and the relative mass $\Delta\ M_{1P}$ of the 1P bottomonium state is the most sensitive to the coupling strength, so we use the measured values of m_π, m_K, and $\Delta\ M_{1P}$ to fix those parameters.

127 There is no really popular-level account of numerical quantum field theory (a.k.a. lattice gauge theory), and probably there never will be. Although some of its results can be described fairly simply, as I've done here, the nitty-gritty is graduate school material. A solid introduction at an unbeatable price can be found here: http://eurograd.physik.uni-tuebingen.de/ARCHIV/Vortrag/Langfeld-latt02.pdf.

Chapter 10

130 **ominous thunderhead:** To minimize the energy the disturbance actually self-organizes into a tube, and the energy is proportional to the length of the tube (as is its mass, according to Einstein's second law). The tube tracks the influence of the quark's color charge, so it can't end (except on an antiquark), and its price in energy is infinite.

Chapter 11

The Physics of Musical Instruments by David Lapp, http://www.tufts.edu/as/wright$_$center/workshops/workshop_archives/physics_2003_wkshp/book/pom_book_acrobat_7.pdf, is a neat, short, mostly non-mathematical introduction to the physics of sound and musical instruments with lots of pictures. Two masterworks are *On the Sensations of Tone* by Hermann Helmholtz (Dover) and *The Theory of Sound* (2 volumes) by Lord Rayleigh (Dover). Only professionals would want to read these tomes from cover to cover, and parts of Helmholtz are obsolete, but just thumbing through them is an inspiration. They'll make you proud to be human.

Chapter 12

136 **Salieri . . . says:** Actually, of course, it's the screenwriter who says these things!

137 **"Actually, Watson . . .":** This joke is adapted from *Quirkology* by Richard Wiseman (Macmillan).

138 **According to an early biographer:** http://www.sonnetusa.com/bio/maxbio.pdf contains *The Life of James Clerk Maxwell, with a selection from his correspondence and occasional writings and a sketch of his contributions to science,* by Lewis Campbell and William Garnett. This is an extraordinary resource on everything Maxwell. In addition to a good old-fashioned biography, it offers an excellent account of his science and a generous selection of his drawings and letters—and even a few of his sonnets.

139 **removing redundant or inessential information:** Several other considerations enter into the design of good data compression, besides the simple goal of keeping the message short. We might want to allow some kinds of mistakes, as long as they don't spoil things too much. JPEG, for example, breaks a continuous image into discrete pixels, and compromises on color accuracy, but usually produces good-looking "reproductions." Or we might want to build some redundancy into the transmitted message, even at the cost of making it longer, if accuracy is precious and the channel is noisy. Reports of measurements from astronomical or GPS satellites get this treatment. Similarly, when people make mathematical models in engineering or economics, for example, they might be very concerned to have equations that are forgiving of errors in manufacture or data, and that accommodate as much empirical input as possible. Theoretical physics, however, puts overwhelming emphasis on compression and accuracy.

141 **for the ultimate in data compression:** On the conceptual foundations of modern data compression, I recommend *Information Theory, Inference, and Learning Algorithms* by David MacKay (Cambridge). For connections to theory-building, and to the work of Gödel and Turing, see *An Introduction to Kolmogorov Complexity and Its Applications* by Ming Li and Paul Vitányi (Springer).

141 **assuming the existence of a new planet:** The history of the discovery of Neptune is complicated and, I understand, somewhat controversial. Alexis Bouvard had already in 1821 suggested that some "dark matter" could be perturbing Uranus (anticipating Bart Simpson). But without a mathematical theory, he could not suggest where to look. John Couch Adams did calcula-

tions suggesting that a new planet could resolve problems with Uranus's orbit in 1843, and supplied coordinates, but he did not publish his work or persuade any observers to follow up.

Chapter 13

146 **namely, as the inverse square:** That is true at macroscopic distances. At ultrashort distances two new effects come into play, and the force laws are different. We've already discussed how Grid fluctuations can modify the force, as the effect of virtual particles diminishes (screens) or enhances (antiscreens) it. Another effect we've discussed is that in quantum mechanics, probing small distances necessarily involves large momenta and energies. This affects the power of gravity, because gravity responds directly to energy. These modifications of the force laws are all-important in the ideas about unification of forces we'll be discussing in Part III.

146 **Gravitational radiation . . . has never yet been detected:** Although gravitational radiation itself has never been detected, one of its consequences has been seen. Precision studies of the binary pulsar PSR 1913+16 over long periods of time indicate that its orbit has been changing in a way that is consistent with the calculated effect of loss of energy due to gravitational radiation. In 1993 Russell Hulse and Joseph Taylor won the Nobel Prize for this work.

Chapter 14

148 **Any body . . . will follow the same path:** Just as an infinite number of straight lines pass through a given point, an infinite number of "straightest" paths, with different slopes, pass through a given point of space-time. They correspond to the trajectories of particles with different starting velocities. So the accurate statement of universality is that bodies starting at the same position *and velocity* will move in the same way under the influence of gravity.

Chapter 16

153 **three lucky people:** David Gross, David Politzer, and I shared the Nobel Prize in 2004 "for the discovery of asymptotic freedom in the strong interaction."

154 **my Italian side:** That's my mom. My father's side is Polish.

154 **"Everything you've said is not even wrong":** The Feynman-Pauli story is well-established folklore in the physics world. I don't know whether it actually happened, and to be honest I don't want to know. It's better left not even wrong.

159 **seed strong force:** The choice of force as a measure of the power of the coupling is somewhat arbitrary. Perhaps a more fundamental measure is the number that appears multiplying the hubs in Feynman graphs, when the process involves particles with the Planck energy and momentum. That number is even closer to unity; it's about 1/2. Any reasonable measure will give a result close to unity—certainly *much* closer than 10^{-40}!

Chapter 17

Ideas to unify the Core by enlarging its local symmetry were pioneered by Jogesh Pati and Abdus Salam and by Howard Georgi and Sheldon Glashow. The *SO*(10) symmetry and classification emphasized in this chapter were first proposed by Georgi. *Grand Unified Theories* by Graham Ross (Westview) and *Unification and Supersymmetry* by Rabindra Mohapatra (Springer) are solid book-length accounts.

166 **It will provide the core . . . possibly forever:** I don't mean to claim that the Core won't be superseded. I hope it will be, and I'm about to describe why and how. But just as Newton's theory of mechanics and gravity remains the description we use for most applications, the Core has such a proven record of success over an enormous range of applications that I can't imagine people will ever want to junk it. I'll go further: I think the Core provides a complete foundation for biology, chemistry, and stellar astrophysics that will never require modification. (Well, "never" is a long time. Let's say for a few billion years.) The Quantum Censor, mentioned in an earlier note, protects these subjects from whatever wild stuff is going on at ultrashort distances and ultrahigh energies.

166 **the weak interaction:** *Weak Interactions of Leptons and Quarks* by Eugene Commins and Philip Bucksbaum (Cambridge) contains extensive discussion of astrophysical applications. *Neutrino Astrophysics* by John Bahcall (Cambridge) is an authoritative treatment by a grand master of the field.

167 **stars live on energy . . . :** The nuclear transformations from which stars derive their energy also include fusion reactions that don't require changing protons into neutrons, such as the process by which three alpha particles (each consisting of two protons and two neutrons) combine into a carbon nucleus (six protons and six neutrons). Such reactions do not involve the weak interaction, but only strong and electromagnetic interactions. They are especially important in the later stages of stellar evolution.

170 **left-handed and right-handed particles:** Really, I should say left-handed and right-handed fields.

A particle with nonzero mass moves at less than the speed of light, and that raises the following problem: You can imagine making a boost so fast as to overtake the particle. To the boosted observer, it will appear to be moving backwards—that is, in the direction opposite to the direction it was moving according to the stationary observer. Because the spin direction still looks the same, a particle that looks right-handed to the stationary observer will look left-handed to the moving observer. But relativity says both observers must see the same laws. Conclusion: the laws can't depend directly on the handedness of particles.

The correct formulation is more subtle. We have quantum fields that create left-handed particles, and separate quantum fields that create right-handed particles. The equations for those underlying fields are different. But once a particle (of either kind) is created, its interactions with the Grid can change its handedness. In the electroweak standard model, interactions of particles with the Higgs condensate do just that.

We can make a strict (that is, boost-invariant) distinction between left- and right-handed for *massless* particles, or using quantum fields. The fact that our successful equations for the weak interactions rely on this distinction shows that Nature prefers massless particles and quantum fields as her primary materials.

176 **The Sirens of myth:** J. Harrison, "The Ker as Siren," *Prolegomena to the Study of Greek Religion* (3rd ed., 1922:197–207) p. 197. This bit is here because I hoped to use *The Siren* by John William Waterhouse as a cover. Alas, it was not to be. But you can see the picture as Color Plate 7.

Chapter 18

Howard Georgi, Helen Quinn, and Steven Weinberg first calculated the behavior of the three forces at short distances, to see if they might unify. (For the strong force, of course, this is just the Gross-Politzer-Wilczek calculation.)

178 **the quantitive measure of their relative power:** Note that at a fundamental level—in terms of the numbers multiplying hubs in Feynman graphs—the weak coupling is actually bigger than the electromagnetic (for experts: here read hypercharge) coupling. Grid superconductivity renders the weak force short-range, however, so its practical effects are usually much smaller.

178 **atomic nuclei . . . are much smaller than atoms:** The contrast between atomic sizes and nuclear sizes is only partly due to the relative weakness of electromagnetic forces. The smallness of the electron's mass, relative to that of protons and neutrons, is also an important factor. We can understand why by recalling the logic of point 3 in the Scholium that concludes Chapter 10. The size of atoms is determined by a compromise between canceling out electric fields by putting electrons right on top of protons, and respecting the wave nature of electrons. The smaller the mass of the particle, the more its wave function wants to spread, and so the small mass of the electron skews the compromise toward larger sizes.

Chapter 19

182 **The famous philosopher Karl Popper:** For much more on Popper and his philosophy, see *The Philosophy of Karl Popper* (2 volumes), ed. P. Schilpp (Open Court).

Chapter 20

The effect of supersymmetry on the evolution of couplings was first considered by Savas Dimopoulos, Stuart Raby, and me. For a personal recollection see Appendix C.

187 **Higgs particles:** You'll find more on Higgs particles in the (popular-level) Oerter and Close books referenced earlier, and more technical discussions in Peskin and Schroeder, and Srednicki.

187 **supersymmetry:** *Supersymmetry: Unveiling the Ultimate Laws of Nature* by Gordon Kane (Perseus) is a popular account by a prominent contributor to the field.

188 **the best idea . . . for connecting them:** Supersymmetry doesn't connect the different parts of the Core directly. None of the presently known particles has the right properties to be the supersymmetric partner of any other. Only by considering both charge unification and supersymmetry simultaneously can we bring everything together.

189 **not *too* much:** SUSY must be broken, but there's even more uncertainty about how this occurs than about the related questions of cosmic superconductivity discussed in Chapter 8 and Appendix B. However supersymmetry breaking occurs, the net result must be that the partners of the particles we know about are significantly heavier. If they're too heavy, they won't contribute enough to the Grid fluctuations, and we'll roll back to the near-miss of Chapter 18.

There are other, independent reasons to suspect that the supersymmetric partners aren't too heavy. The most important one is this:

If you calculate the effect of virtual particles on the mass of the Higgs particle in a unified theory, you find that they tend to pull that mass up to the unification scale. This is the essence of what's often called the hierarchy problem. You can cancel off these effects with the stroke of a pen by having the starting mass be just enough to cancel the virtual-particle contributions almost precisely, but most physicists find such "fine-tuning" repellant—they call it *unnatural.* With supersymmetry the corrections cancel, and not so much fine-tuning is required. But if supersymmetry is badly broken—i.e., if the partners are too heavy—we're back in the soup.

189 **Now we must correct the corrections:** In this calculation, I include the effects of only the particles necessary to implement supersymmetry. (For experts: I'm dealing with the minimal supersymmetric standard model, or MSSM.)

Additional (much heavier) particles necessary to make a complete unified theory have not been included. That is why the couplings, after unifying at high energies, diverge again. In the full theory, once they come together they will stay together. But since we don't know enough to pin down relevant details of the full theory, I opted to take things as they come.

191 **they all come together, pretty nearly:** Since we don't have a reliable theory of how gravity behaves at short distances I've just sketched in the gravity line roughly.

Chapter 21

For more information on the LHC project, including the latest news, you can visit the CERN website http://public.web.cern.ch/Public/Welcome.html and follow links from there. *Perspectives on LHC Physics*, edited by G. Kane (World Scientific), is a collection of articles by leading experts. I also recommend my scientific paper "Anticipating a New Golden Age." You can find it at itsfrombits.com.

196 **The dark-matter problem:** *Quintessence* by Lawrence Krauss (Perseus) is a good popular account of the dark matter, dark energy, and modern cosmology in general.

198 **protons should decay:** Its insensitivity to details is both the strength and the weakness of our calculation showing that (low-energy) SUSY provides accurate unification of the forces. A new particle's contribution screening (or antiscreening) kicks in only at energies larger than the rest energy mc^2 of the particle. Because the changes crucial to unification accumulate over a vast span of energies, it doesn't much matter exactly where they begin, so the particle's contribution depends only mildly on its mass. Thus the result of our unification calculation wouldn't be much affected if the masses of the new SUSY particles were (say) to double or to halve. The result is stiff: it can't be bent easily. Proton decay, however, does depend on the details.

198 **what new effects to expect:** String theory has inspired speculation about the existence of extra spatial dimensions. The extra dimensions must be either very small (folded up) or highly curved and difficult to penetrate; otherwise we'd have noticed them. Maybe a closer look, using the LHC, will reveal them. *Hiding in the Mirror* by Lawrence Krauss (Viking) and *Warped Passages* by Lisa Randall (Harper Perennial) give popular accounts of these ideas.

Epilogue

202 **that's a far cry from explaining:** According to the ideas described in Appendix B, the Higgs condensate is directly responsible for the mass of W and Z bosons, through a form of cosmic superconductivity. So if those ideas are

right, once we figure out what the Higgs condensate is, we'll understand the origin of those particular masses.

Appendix B

215 **So we can easily generate . . . the decay:** You can check that by emitting a boson that carries off a unit of red charge and brings in a unit of purple charge (that is, *carries off* a *negative* unit of purple charge), the u quark in the top line will change into the antielectron e^c on the fifteenth line. (The + and – entries are *half*-unit charges.) By absorbing the same boson, the d quark on the fifth line will change into the anti-u quark on the ninth line. In the Charge Account, these guys are related by flipping + and – signs between the first and last columns. So by emitting and absorbing this particular color-changing boson, as a virtual particle, we get the process

$$u + d \rightarrow u^c + e^c$$

Now we add a spectator u quark to both sides: $u + u + d \rightarrow u + u^c + e^c$, and we're almost home. We just have to recognize that the $u + u + d$ are the ingredients of a proton, and that $u + u^c$ can annihilate into a photon. Finally then we arrive at the proton decay process

$$p \rightarrow \gamma + e^c$$

as promised.

Illustration Credits

Figures

Figure 7.3 is based on a drawing that first appeared in my article "QCD Made Simple" in *Physics Today*, 53N8 22–28 (2000), and is used with permission.

Figure 8.1 is based on an illustration that appeared in documentation for the ArcInfo Workstation from ESRI, and is used with permission.

Figures 9.1, 9.2, and 9.3 are based on an updated version of work reported by the MILC collaboration, which appeared in *Physical Review* D70, 094505 (2004) (Figure 17), and is used with their permission.

Color Plates

Plates 1, 2, 8, and 9 are taken from the CERN image library, and are used with permission.

Plate 3a is based on the painting *The Library of a Mathematician* by Aldo Spizzichino, and is used with his permission.

Plate 3b is based on the work of Michael Rossman and his group, reported in *Nature* 317, 145–153 (1985), and is used with his permission.

Plate 4 is due to Derek Leinweber, CSSM, University of Adelaide, and is used with his permission.

Plate 5 is based on work of the STAR collaboration, and is reproduced Courtesy of Brookhaven National Laboratory.

Plate 6 is based on work of Greg Kilcup and his group, and is used with his permission.

Plate 7 is a reproduction of the painting *The Siren* by John William Waterhouse (circa 1900).

Plate 10 is due to Richard Mushotzky, and is used with his permission.

Appendix C

Appendix C is based on my article of the same name that first appeared in *Nature* 428, 261 (2004), and is used with permission.

Index

PENGUIN SCIENCE

IN SEARCH OF THE MULTIVERSE
JOHN GRIBBIN

In Search of The Multiverse takes us on an extraordinary journey, examining the most fundamental questions in science. What are the boundaries of our Universe? Can there be different physical laws to the ones we know? Are there in fact other universes? Do we really live in a Multiverse?

This book is a search – the ultimate search – exploring the frontiers of reality. Ideas that were once science fiction have now come to dominate modern physics. And, as John Gribbin shows, there is increasing evidence that there really is more to the Universe than we can see. Gribbin guides us through the different competing theories revealing what they have in common and what we can come to expect.

Along the way Gribbin explores the very latest thinking about quantum theory, about gravity and the fundamental forces that shape our world, about time and multiple dimensions, about matter itself, and the growth and fate of the known Universe.

John Gribbin is our best, most accessible guide to the big questions of science. And there is no bigger question than our search for the Multiverse.

read more

PENGUIN SCIENCE

SOME TIME WITH FEYNMAN
LEONARD MLODINOW

In Mlodinow's intimate portrait of Feynman we look into his unfolding thoughts on the nature of creativity, his rivalry with colleague Murray Gell-Mann, his love for the women in his life, and the cancer that would ultimately kill him.

The result is a fascinating picture of an irreverent, charismatic and startlingly honest man, who believed in taking risks and breaking rules and who did research not from ambition, but for the thrill of discovery.

'An accessible picture of a brilliant man' Stephen Hawking

'A warm, delightful glimpse into a fascinating world' *Scotland on Sunday*

SIX EASY PIECES
RICHARD P. FEYNMAN

Drawn from Richard Feynman's celebrated and landmark text 'Lecture on Physics', this collection of essays reveals Feynman's distinctive style while introducing the essentials of physics to the general reader.

'A delightful volume – it serves as both a primer on physics for non-scientists and as a primer on Feynman himself ' Paul Davies

THE ORIGINS OF VIRTUE
MATT RIDLEY

'Are we driven by a profoundly selfish, determinist impulse? Or is there an escape clause that enables us to be genuinely unselfish and good? In an era in which biological science is challenging traditional ethics, he has raised the debate to a new level of seriousness and importance' *Sunday Times*

'A brilliant, lucid insight into the profound implications of modern biological thinking' Bryan Appleyard

PENGUIN SCIENCE

PENGUIN SCIENCE

THE BLANK SLATE STEVEN PINKER

'The best book on human nature that I or anyone else will ever read. Truly magnificent' Matt Ridley, *Sunday Telegraph*

'A passionate defence of the enduring power of human nature … both life-affirming and deeply satisfying' Tim Lott, *Daily Telegraph*

'Brilliant … enjoyable, informative, clear, humane' *New Scientist*

'If you think the nature/nurture debate has been resolved, you are wrong. It is about to be reignited with a vengeance … this book is required reading' *Literary Review*

'Startling … Pinker makes his main argument persuasively and with great verve … This is a breath of air for a topic that has been politicized for too long' *Economist*

HOW THE MIND WORKS STEVEN PINKER

'Why do memories fade? Why do we lose our tempers? Why do fools fall in love? Pinker's objective in this erudite account is to explore the nature and history of the human mind' *Sunday Times*

'Witty popular science that you enjoy reading for the writing as well as for the science' *The New York Review of Books*

THE LANGUAGE INSTINCT STEVEN PINKER

'A marvellously readable book…illuminates every facet of human language: its biological origin, its uniqueness to humanity, its acquisition by children, its grammatical structure, the production and perception of speech, the pathology of language disorders and its unstoppable evolution' *Nature*

'An extremely valuable book, informative and well written' Noam Chomsky

'Brilliant … Pinker describes every aspect of language, from the resolution of ambiguity to the way speech evolved … he expounds difficult ideas with clarity, wit and polish' Stuart Sutherland, *Observer*

He just wanted a decent book to read ...

Not too much to ask, is it? It was in 1935 when Allen Lane, Managing Director of Bodley Head Publishers, stood on a platform at Exeter railway station looking for something good to read on his journey back to London. His choice was limited to popular magazines and poor-quality paperbacks – the same choice faced every day by the vast majority of readers, few of whom could afford hardbacks. Lane's disappointment and subsequent anger at the range of books generally available led him to found a company – and change the world.

'We believed in the existence in this country of a vast reading public for intelligent books at a low price, and staked everything on it'
Sir Allen Lane, 1902–1970, founder of Penguin Books

The quality paperback had arrived – and not just in bookshops. Lane was adamant that his Penguins should appear in chain stores and tobacconists, and should cost no more than a packet of cigarettes.

Reading habits (and cigarette prices) have changed since 1935, but Penguin still believes in publishing the best books for everybody to enjoy. We still believe that good design costs no more than bad design, and we still believe that quality books published passionately and responsibly make the world a better place.

So wherever you see the little bird – whether it's on a piece of prize-winning literary fiction or a celebrity autobiography, political tour de force or historical masterpiece, a serial-killer thriller, reference book, world classic or a piece of pure escapism – you can bet that it represents the very best that the genre has to offer.

Whatever you like to read – trust Penguin.